MADE IN BRUNEL™
Making Our Mark

MADE IN BRUNEL
Brunel University, Uxbridge, UB8 3PH
01895 267776
www.madeinbrunel.com
info@madeinbrunel.com

Editors: Jordan Chitty, Simon Gilbert,
Jonathan Marsh, Helen Wright,
Stephen Green, Paul Turnock

Typeset in Frutiger 45 Light

First published in 2007 in collaboration with
Papadakis Publisher

An imprint of New Architecture Group Ltd
11 Shepherd Market, London W1J 7PG
www.papadakis.net

ISBN: 978-1-901092-85-1

Printed and bound by Nuffield Press in the
United Kingdom

FOREWORD

It is a great pleasure to be asked to introduce this excellent piece of work, filled with fantastic, innovative ideas from some of the brightest young minds in the country.

The creativity showcased through the MADE IN BRUNEL brand is inspiring, but so too is the degree to which these projects represent genuine commercial opportunities and I am impressed with the business focus displayed by many of these students.

I am confident that the MADE IN BRUNEL students will contribute decisively to the future success of the British economy and I very much look forward to seeing their careers take off.

Sir Digby Jones

CONTENTS

PREFACE

It is amazing what we can do when we put our minds to it. We have achieved so much during our lives here at Brunel. We really have made our mark, not only in a physical sense, but spiritually, and emotionally upon friends and colleagues we have met along our rollercoaster journey.

MADE IN BRUNEL connects us all in ways that we could not have thought when we first joined the project. We see it as a means to display our strengths, our pathway to career success and this year is an opportunity to make our individual mark; to individualise the impact that we have, far and wide.

Making Our Mark does not stop here. As you will see throughout this book, Brunel students are not only proud of their accomplishments, but also brim with enthusiasm about the future and the many possibilities it holds for us all.

We look forward to every mark that we will make in our careers; whether it is something to make you smile, help you out, take you on a journey, or save your life.

Whatever it is, that is our goal.

Simon Gilbert and Jordan Chitty
MADE IN BRUNEL Project Coordinators 2007

Her Majesty The Queen visiting Brunel, as part of university celebrations to mark the bicentenary of Isambard Kingdom Brunel, the university's namesake, and also the 40th anniversary of the university's Royal Charter.

As part of the visit, on 19 May 2006, The Queen met students from the School of Engineering and Design, discussing with them the projects that they worked on throughout their time at Brunel.

MAKING OUR MARK

Creative, innovative and entrepreneurial young people are the fuel for the UK's thriving economy. Engineers and designers create solutions that become new services and businesses, not just products. All kinds of industries, from biotechnology through energy to financial services, benefit from people who can think out of the box while delivering outcomes that are practical.

There is a significant and increasing contribution to business from design and technology with great support from the Government. The challenge for us all, especially in response to factors such as climate change, health and globalisation, is to improve and leverage our use of technology and design through innovation and creativity.

At MADE IN BRUNEL you can see the exciting results of a dynamic, disciplined process of innovation. The graduating students don't just bring their technical achievements and creativity to the show but their understanding of markets, communications skills, ability to manage projects to tight deadlines and budget constraints and, above all, their passion for what they do.

These students come from a range of backgrounds and cultures. Brunel offers a unique combination of diversity; a creative, competitive environment disciplined by academic and technical standards of excellence, with inspirational teachers. Somehow this mix, in the innovation cluster of London, produces graduates with such energy and creative drive, that they refute the clichés that creativity and entrepreneurship cannot be taught.

These young people, and the companies that they create or which sponsor or employ them, will make their mark in our lives, our economies and in the wider world. Some of them will go on to contribute to solutions to some of the worlds' most difficult problems.

I believe that MADE IN BRUNEL is not just about showing off what our students and graduates can do, but also about attracting other young people to engineering and design. It is about inspiring businesses to innovate, to invest in new ideas and it is about reaching out to a wide range of people with the sheer fun of innovation.

Working with the team and the wonderful people who have supported, sponsored and inspired MADE IN BRUNEL has been a great privilege. I particularly want to thank Simon Gilbert, Brian Kingham, Dame Mary Richardson, Paul Turnock, Joseph Giacomin, Paul Sinclair and Stephen Green for making this all possible.

Lady Chisholm, MA Cantab., BSc (OU), FRSA.
Trustee, NMSI and REACH.
Advisory Board member JRBH Strategy
and Management and OBHE

INNOVATION AS PARADIGM:
THE ROLE OF MADE IN BRUNEL

A quick glimpse at the newspapers, or the odd minute or two in front of the television, quickly fills the eyes and ears with great and exciting stories of our changing economic landscape, of globalisation, of new complexities, of multiethnicity and of opportunity. Gone appear to be the days of the endless rows of factories with their smoking chimneys, dangerous machines and crowded human masses. The new century has brought a more diverse economic landscape, something of a return to the past, with the individual-artisan benefiting from numerous and profitable opportunities for fulfilling her or his human potential as part of an economic network which spans the now small globe. The designer in London is sending her sketches to her friend, the prototyper in Italy, who is sending production drawings to his contact, the manufacturer in Malaysia, who is sending his production batch to a distributor in Russia, who is offering the final packaged product to the European public. The world is small, the information is great, and the situation is summarised for many people by the buzzwords 'innovation' and 'creative industries'.

The exciting new world demands, however, changes of its inhabitants. As has always been the case since man first developed tools, rigour and discipline remain important. A new requirement of today's society, however, which separates it from those of our ancestors is the perceivable, touchable, unavoidable urgency of personal creativity. Willingness, energy and fantasy are required to push the boundaries and to reap the benefits of our new economic landscape. It is within the framework of this exciting new world that MADE IN BRUNEL can be analysed, understood and appreciated. What is the role of an international centre of education in this new landscape? Is it to provide skills then withdraw? Is it to provide wisdom then let knowledge takes its course? Is it to help those who aspire to a profession to find a job? Arguably, it is all of these, and more.

The contents of this directory can be understood as the final step in the preparation for a career in today's creative industries. The aspiring creative industrialists whose work is presented within, have integrated their personal creativity, their recently acquired skills and their willingness to interact with the global economic system into projects. These have two faces: that of the individual's wish to join the global creative industry, and that of the their wish to change the global creative industry. Both conformity and rebellion fill the pages, a will to create and a will to destroy. Innovation in its simplest and purest form.

From healthcare products to consumer goods, from industrial machines to art, the MADE IN BRUNEL collection oozes innovation. With MADE IN BRUNEL the university has achieved a passage through which aspiring talents enter the new economic landscape, and a platform for those who already happily inhabit the landscape from which to review, enjoy and digest what our new colleagues think, feel and want. As a leading international centre of education, the university is proud to provide this important opportunity to the new talents, but even more proud of providing a window into the world of tomorrow for those with the foresight to engage with the new generation of creative industrialists. Innovation is the paradigm, MADE IN BRUNEL is the manifestation.

Professor Joseph Giacomin
Director of Human Centred Design Institute
Brunel University

The MADE IN BRUNEL team 2007

Project Coordinators
Simon Gilbert
Jordan Chitty

Creative Team
Helen Wright (Team Leader)
Sam Bairstow
Jay Canham
Jennifer Cheung
Maria Dolka
Owen Glynn
Stefan Grosvenor
Jonathan Marsh
Heena Rai
Paul Tinker
Sam Weller

Digital Team
David Strugnell (Team Leader)
Steven Boulton
Nadia Hussain

Ghaalib Khan
Daniel Noonan
Yone Santana
John Stewart
Duncan Stevenson

Pecha Kucha Team
Matthew Higham
Jade Hutchinson

PR and Marketing Team
David Gadd (Team Leader)
Carlo Belli
Robert Bruns
David Connell
Ben Griffiths
Malak Kamil
Thomas Pynn
Sophia Richards
Charlie Spencer

Show Team
Laura Baird (Team Leader)
Matthew Barnett
Julian Charity
Jeremy Crouch
Simon Elliott
Alexander Hill
Fenella Holden
Mark Jones
Sophia Kelly
Jeffrey Knapman
Thomas Leech
Louise McKillop
Chris Miniken
Will Postle
Stephanie Prichard
Barney Stephens
Lindsey Stevenson
Laura Williams
Natalie Vanns
Meghana Vaidyanathan

HSBC GLOBAL EDUCATION TRUST AND MADE IN BRUNEL

The MADE IN BRUNEL exhibition encourages the development of design, engineering and enterprise skills in an exciting and original way. HSBC Global Education Trust is delighted to be supporting, for a second year, this showcase of some of the most innovative design and engineering ideas from exceptional students at the beginning of their careers.

We welcome the opportunity to facilitate the visit of staff and students of Tsinghua University, China, and IIT (Indian Institute of Technology) Madras, India. International cooperation and trust are encouraged through greater understanding of other cultures. This event is an opportunity for young people to exchange ideas and learn about new concepts from the most gifted students across the world.

Dame Mary Richardson
Chief Executive of the HSBC Global Education Trust

PRODUCT INNOVATION

New ideas allow the progressive
change of our living environment and
reflect our multi-faceted cultures. The
development of any new product for
people requires intellect, imagination
and technical ability.

ANDREW BROWN

INDUSTRIAL DESIGN BSc

Product Innovation

I have enjoyed my time at Brunel and during the four years my interest in design has grown considerably and I have made many good friends along the way. I feel that the knowledge and skills gained along with increased self-confidence will aid me in whatever I attempt in the future.

Skills: Adobe Photoshop, Illustrator, InDesign and Dreamweaver; SolidWorks, AutoCAD, 3D Studio Max, V-ray, Pro/ENGINEER.

Experience: 2005 – 2006 Futurama Ltd., West Molesey

Interests: music, playing guitar, football, graphic design, photography, motor racing and films

T   +44 (0)7811 900073
E   andrewbrown85@gmail.com

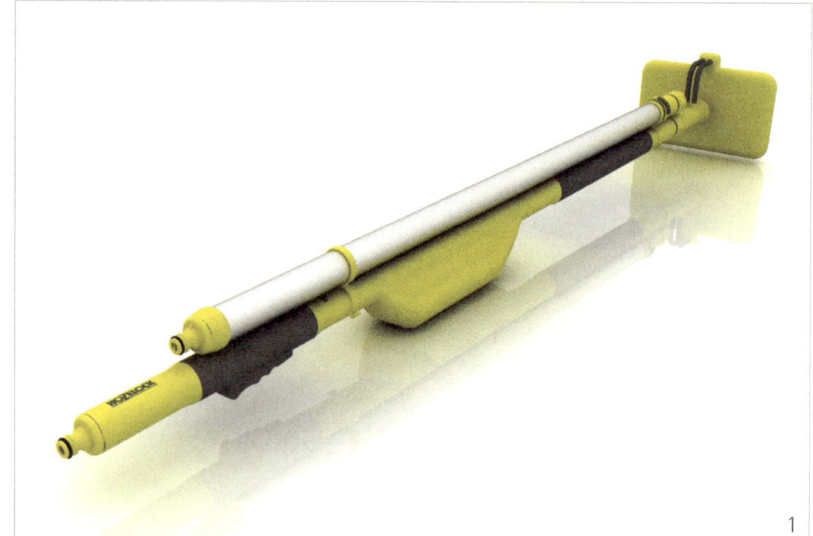

1 Deionising window cleaner that utilises the process of deionisation of water to remove salts and impurities, leaving a streak-free clean. This allows the user a quick and easy alternative to paying for window cleaners. The product was fully modelled and rendered using 3D Studio Max and V-ray including all internal tubing, components and connectors. 2 Photography is of great interest to me and it is something that I wish to develop further after leaving Brunel.

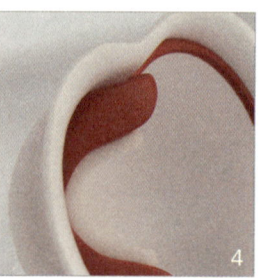

3, 4 Yakult Breather is a breathing mask that can be worn in cities or areas with high levels of pollution, the Breather acts as a filter so that the wearer only breathes in clean, fresh air keeping to the Yakult ideology of staying healthy. 5 Graphic design has become my favourite aspect of design and I take great joy in experimenting with new techniques and ideas to create interesting and innovative pieces of graphic art.

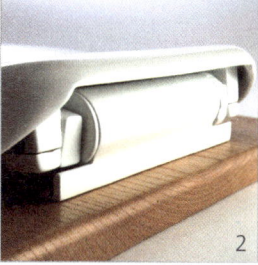

ROBERT BRUNS

INDUSTRIAL DESIGN BSc

- Product Innovation -

Self-closing Toilet Seat. The toilet lid or seat closes after a time delay, only once the user has finished. The solution is entirely free of electronics, utilising a viscous fluid with a change in pressure to provide the delay. Market research indicated that the product should be available for under £25, and this is achieved through efficient material usage. 1 Working prototype. 2 Aesthetic model, close up of mechanism. 3 Working prototype, delay and damping mechanism. 4 Aesthetic model.

A placement year at Hasbro Europe has helped me to develop invaluable skills in design for manufacture and to create products which reflect the core brand values of the company. My work achieves function through simplification with the belief that function and form are not mutually exclusive. I strive to add value to products through the introduction of intelligent design. I hope to strengthen these skills in the future, within an enthusiastic working environment.

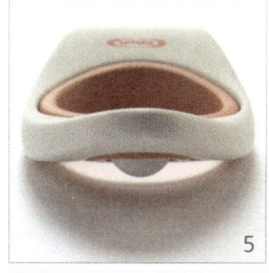

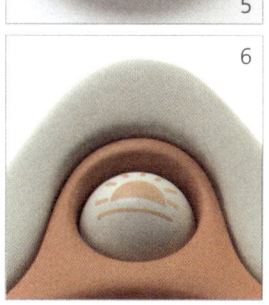

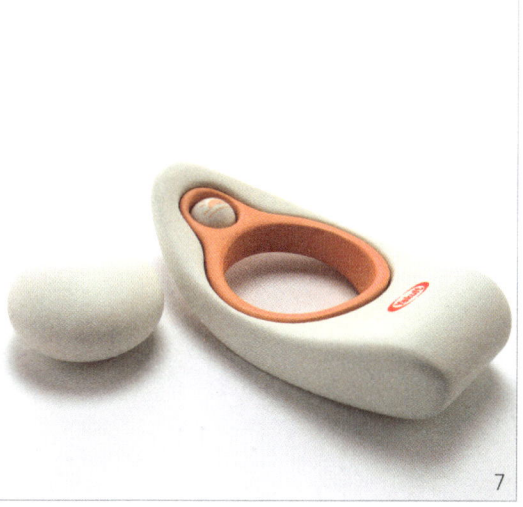

5 – 7 Sleep. A conceptual product for Yakult, to achieve emotional reassurance. 'Sleep well, live well.' The product dock slowly stands up throughout the night, allowing the user to gauge how much sleep time remains. To create an emotional connection, the pebble like unit is taken to bed, and gently awakens the user in the morning when the dock is fully upright. 8 Travel Downfall, a children's game currently in production, restyled whilst on placement at Hasbro Europe. (Previous version in background).

T   +44 (0)7758 266828
E   robert.w.bruns@gmail.com

JORDAN CHITTY

INDUSTRIAL DESIGN BSc

Product Innovation

I strongly believe that good design not only results in better products, but a better business too.

A year working at leading design consultancy IDC followed by coordinating the MADE IN BRUNEL team have given me capabilities beyond those taught on the course. My graphic and website design business targets all of my skills towards helping companies large and small improve their bottom line through investing in design.

T   +44 (0)7792 791009
E   jordan@jordanchitty.com
W   www.jordanchitty.com

Heinz Digital Guardian protects your family's digital memories. In addition to its anti-virus and firewall system, it backs up all your data and features 'HeinzSight' to retrieve lost files.

Tank2Go is a response to a client brief to develop an Integrated trolley system for scuba diving twin sets. Designed as an upgrade to any twin set, it remains on the tanks when diving.

RUPERT DAVIES

INDUSTRIAL DESIGN BSc

Garden GT – a streamlined, adaptable garden workstation with superior refuse capacity. Refuse bucket pivots for ease of emptying and is detachable for maximal base unit loading. It also features a rear storage facility for gardening paraphernalia.

Experience: 12 month placement with Ordnance Survey working as a Graphic Designer.

Skills: Adobe InDesign, Photoshop and Illustrator, Autodesk 3ds Max and AutoCAD, Pro/ENGINEER, freehand drawing and workshop proficiency. Interests: music, drumming, film, technology, snowboarding, surfing and the great outdoors.

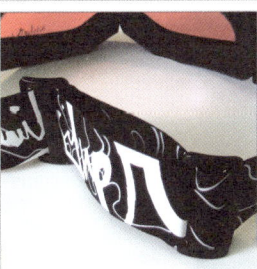

CONTOUR – futuristic concept for the Quiksilver Sixth Sense product range. These vision-enhancing goggles map out the piste and detect obstacles in low visibility. Data is translated and relayed to a HUD projected onto the interior lens.

T   +44 (0)7973128452
E   ruedavies@gmail.com

BEN DRUETT

INDUSTRIAL DESIGN AND TECHNOLOGY BA

— Product Innovation —

Experience: Twelve month placement working with Knorr-Bremse Rail Systems Ltd.

Skills: Pro/ENGINEER Adobe Photoshop, Illustrator, InDesign, Microchip PIC16F MCU

T   +44 (0)7814 357032
E   bendruett@hotmail.com

Home Micro Brewery Improved efficiency of traditional brewing techniques allows the user to brew 40 pints of beer easily and economically. A variety of different ingredients can be used in the micro brewery to customise the beer produced. The process is controlled from an electronic control box where the temperature of the water can be monitored on an LCD Screen.

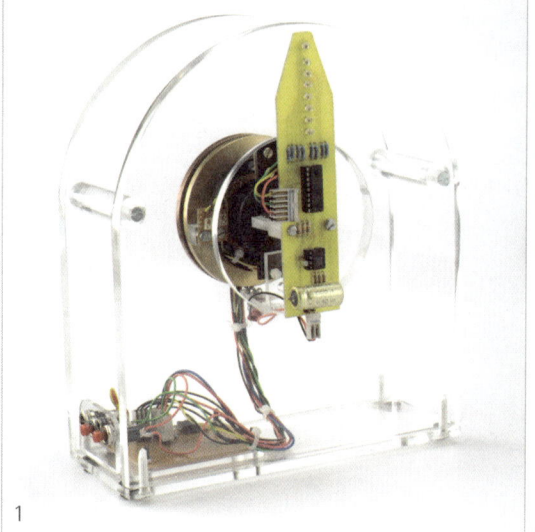

1

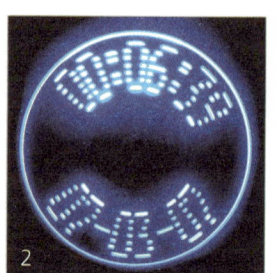

2

3

1, 2 Propeller Clock Displays the time and date with just 7 LEDs As the propeller spins, the LEDs turn on and off in a sequence to display the image. Data is communicated between the base and the propeller via infra red sensors. Power is transmitted to the propeller via slip rings. 3, 4 North Face Cycle computer - Integrated lights, rear camera and LCD screen enable the user to clearly see potential hazards behind them.

4

BEN GRIFFITHS

INDUSTRIAL DESIGN
AND TECHNOLOGY BA

- Product Innovation -

1 – 4 The SteamMachine is a self-guiding buggy, utilising sonar and reversing sensors to judge distances, enabling the buggy to avoid obstacles and navigate within confined spaces. The buggy is powered using a twin, double-acting, reversible steam engine. The vehicle is low profile and has monster truck wheels. This combined with a mix of brushed aluminium and brass forms a distinctive "steam punk" style.

The culmination of skills and knowledge gained over the last four years at Brunel has led me to believe that there are few limits to what we as designers can achieve. An exchange programme to the University of Technology Sydney, coupled with an industrial placement at Wall Colmonoy has also provided me with a greater understanding of manufacturing techniques and different cultural approaches to design.

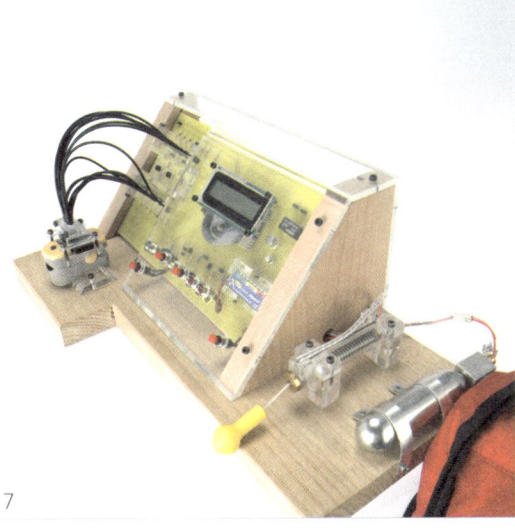

5, 6 Marmite jewellery incorporating a GPS system to provide the wearer with a map of their day.
7 The smart wetsuit incorporates a spectrophotometric system to monitor the user's consciousness level, triggering an inflation system if these levels drop critically, preventing the user from drowning.

T   +44 (0)7787 130613
E   bengriff101@hotmail.com
W   bengriffiths.net

DAN HARVEY

INDUSTRIAL DESIGN BSc

Product Innovation

A good designer is a problem solver with the right question in mind; asking not what, but why? I consider myself a problem solver, an artistic engineer who heeds criticism instead of rejecting it. Brunel University has helped mould me into a well developed designer. Spending a year at San Francisco State University allowed me to broaden and diversify my design skills and abilities. I am fluent in the following programs: SolidWorks, Pro/ENGINEER, AutoCAD, AliasStudio, Adobe Photoshop, Illustrator, InDesign, Flash and Dreamweaver.

T   +44 (0)7709141296
E   dannyjharvey@hotmail.com

Quiksilver BLAZE is a future design concept for a technologically advanced skateboard truck developed for the Quiksilver Sixth Sense product range. The skateboard truck emits a UV ink upon performing a 'grind', leaving a trail as it goes. This trail of success is seen under UV lighting but is invisible in normal lighting conditions; so no traces of the ink are visible in daylight.

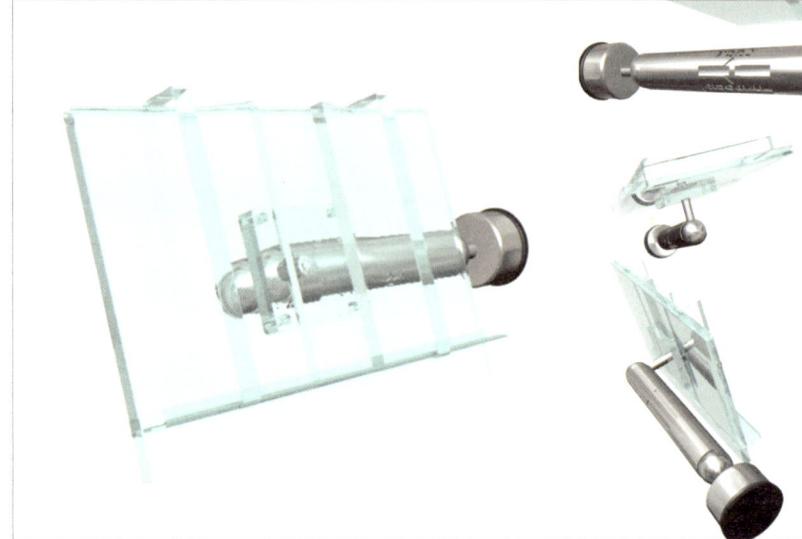

Bathroom reading stand that allows users to read in the bathroom in comfort. This fully adjustable reading stand is capable of holding a diverse range of reading material. The stand holds pages open without interfering with the users reading capability. The style and finish of the product has been chosen in order to form a desirable specialist product, which would be a stylish addition to any luxurious bathroom.

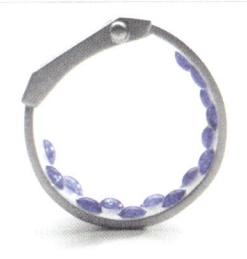
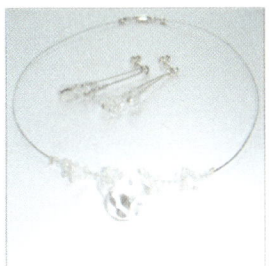

DAISY HODGSON

INDUSTRIAL DESIGN BSc

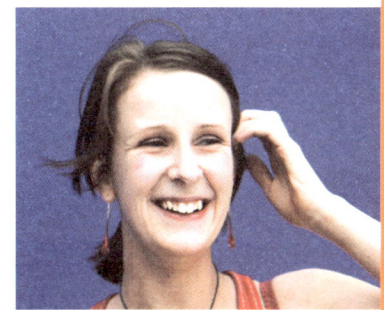

My time at Brunel has not only helped me to gain a good grounding in all aspects of design, but also through my placement at Harriet Kelsall Jewellery Design I have found the particular area of design that I wish to explore in a future career.

Silver ring with Jadeite stone in all around setting made on placement at Harriet Kelsall Jewellery Design. Side view of watch designed for Red Bull. Rock crystal necklace and earring set designed for Harriet Kelsall Jewellery Design. Watch designed for Red Bull that gives children confidence when away from their parents.

T   +44 (0)7732 041035
E   thegrrnoise@hotmail.co.uk

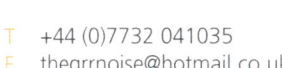

MARK JONES

INDUSTRIAL DESIGN
AND TECHNOLOGY BA

Working for Redten Digital as their in-house designer developed my skills in drawing metalwork and parts using SolidWorks for external fabrication. Talking to customers and designing their ideas into products from a prototype stage to manufacture greatly improved my communication skills. The Digital photo kiosks that I worked on as part of my experience are now in use at Tesco and Klick Photopoint stores. It was a great year working within a team of dedicated people motivated to move the company forward.

T   +44 (0)7739 838799
E   mjones_uk2@hotmail.com

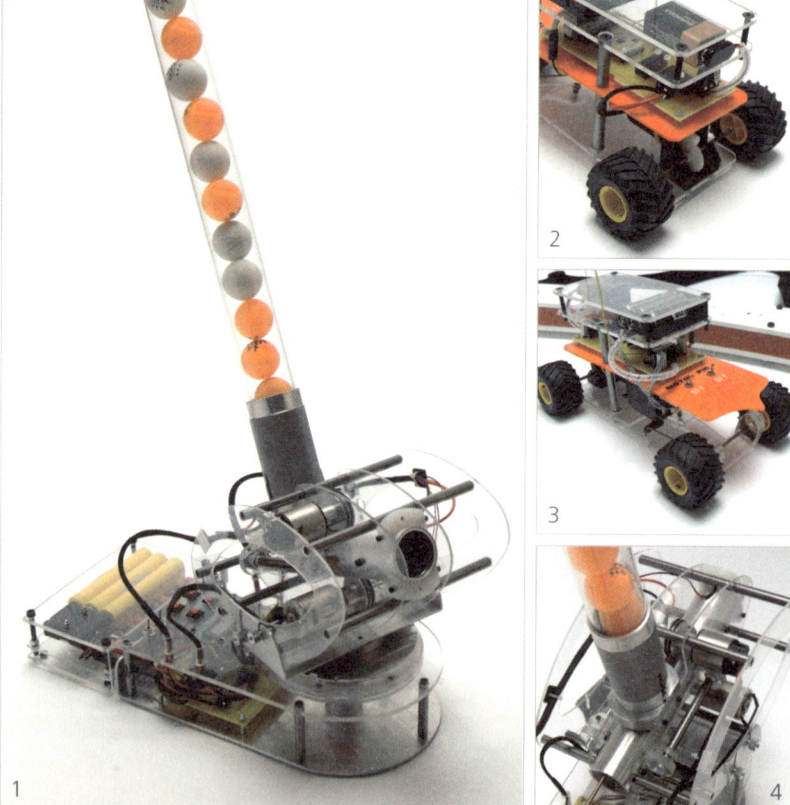

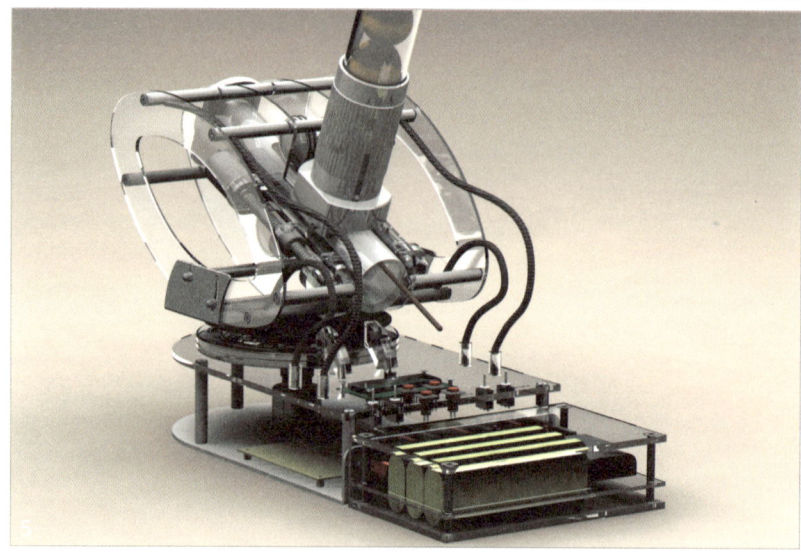

1, 4 – 5 'TT23' – A personal table tennis trainer to improve the development of novice players, producing topspin, backspin and 18 training routines. 2 – 3 'Handy Car' A small toy car controlled wirelessly by the movement of hands without having to touch any controls. Moving the right hand vertically controls the speed and lateral movement of the left hand controls the steering.

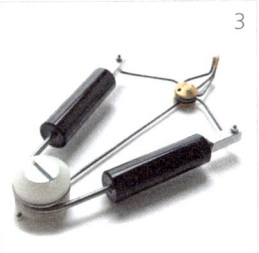

1 Let Loose De-braider, aesthetics model. 2 Let Loose Logo. 3 Let Loose De-braider, working prototype. It releases human hair braids 60% faster than using hands or a comb and is 30% faster with synthetic hair braids. 4 Coconut Chilli logo, Indian Cuisine.

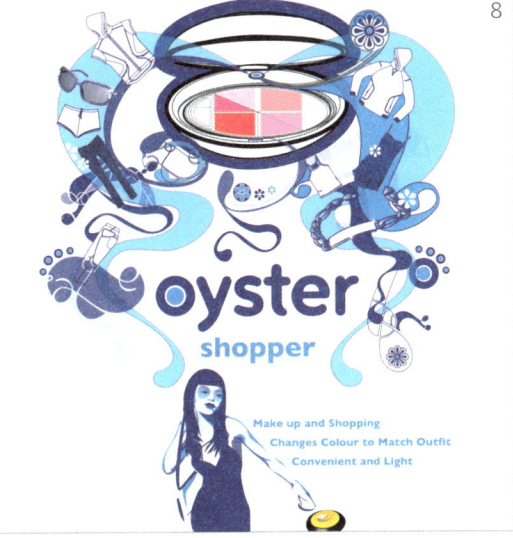

5 Oyster Shopper, aesthetics model. Future concept for Oyster that allows users to shop, pay and apply make up in one pocket-sized product. Other benefits include trying clothes on, locating and holding items on the touch screen. 6 Oyster Shopper, closed with packaging. 7 Painting, acrylic paint. 8 Oyster Shopper Adcept (Advertising concept).

MALAK KAMIL

INDUSTRIAL DESIGN BSc

▪ Product Innovation ▪

T   +44 (0)7985 193058
E   malak59@brunel.ac.uk

Drawing, painting and sketching all contribute to my love of graphic design. Creating artwork to please the eye is my passion and I am versatile with applying these skills throughout the design process from branding to product styling. Studying design at Brunel has given me a large insight into the challenges designers can face and has helped me become an efficient problem solver.

Experience: 2005-2006 Monkeehouse Ltd, Creative Designer. Bionyc Industries, Creative Designer. Crush Licensing, Portfolio Manager. 2006-2007 MADE IN BRUNEL Marketing Team Member

Skills: Adobe Illustrator, Photoshop, InDesign, Pro/ENGINEER, metal/wood/plastic workshop skills, modelmaking, prototyping, painting and drawing.

MOE KRIMAT

INDUSTRIAL DESIGN BSc

The wide range of skills and priceless work experience gained at Brunel has given me a taste for versatility and professionalism that I seek to broaden. It is in the understanding of a complex matter that I seek to retrieve the ulterior motive. A brief comes as a collision of infinite images and words. It is only in the final process of conveying the subliminal message in a public medium that a designer is seen in his true light.

Oakley Tuatara provides the technology required for an athlete to compete against the merciless forces of the desert. With a combination of ventilation tunnels lined with heat sinks and intelligent fabric to help increase perspiration, the Tuatara head gear effectively reduces the chances of hyperthermia when sandboarding

T   +44 (0)7748 976409
E   krimatm@hotmail.co.uk

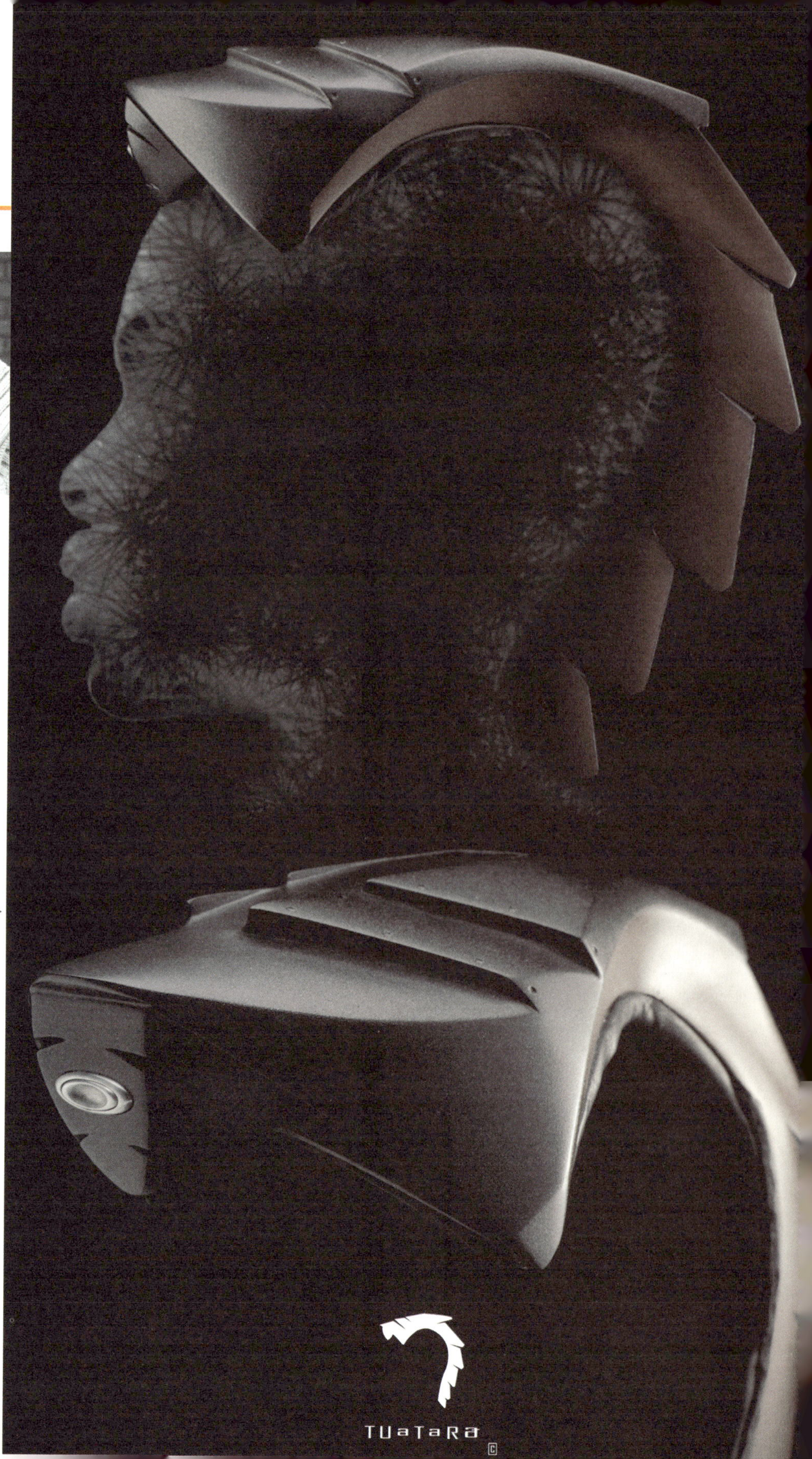

TUaTaRa

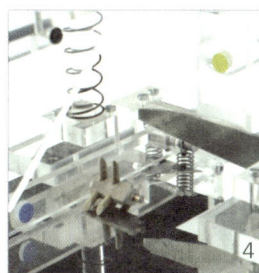

JOHN MAHER

INDUSTRIAL DESIGN BSc

— Product Innovation —

Paper Fastener A paper fastening method designed to meet paper use requirements without the need for refills. 1, 4 Prototype applicator mechanism. 2 Applicator form. 3 Fastening applied to a collection of sheets.

Time I have spent both studying and on placement in industry has fostered an enjoyment of a hands-on approach to problem solving and the design process.

Experience: 14 months industrial placement at Icore International working as part of the development team on the design and development of new products for electrical and aerospace applications.

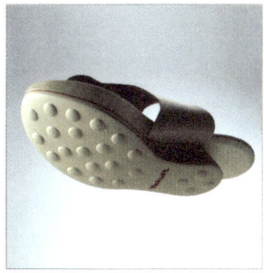

Yakult Footwear A concept developed for the Yakult brand, Yakult Footwear helps the body to walk in a more natural way. The human musculoskeletal system is designed to walk barefoot on soft, natural ground. Yakult Footwear replicates this by encouraging the wearer to work core muscles whilst walking, bringing health benefits to the whole body.

T   +44 (0)7891 717043
E   j.p.maher@hotmail.co.uk

JONATHAN MARSH

INDUSTRIAL DESIGN BSc

Product Innovation

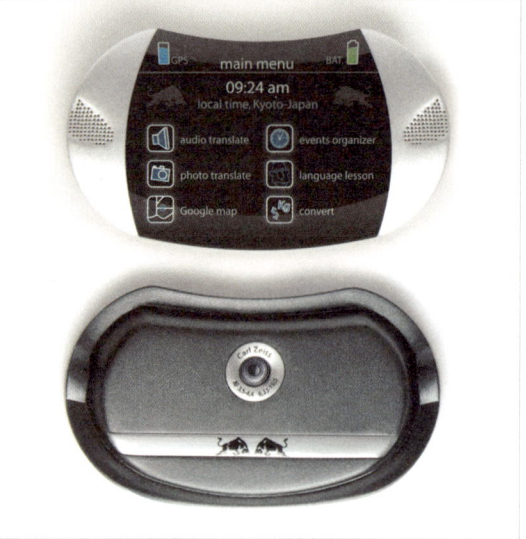

Design, in the modern sense of the word, is no single entity, but rather a bespoke series of processes that are required to develop the best possible solution to a problem. In this way designers must constantly adapt and learn new processes and technologies to deal with this diversity. At the same time the consumer is becoming ever more knowledgeable about the products they buy. Consumers are looking for perfection in all areas of a product and designers must satisfy that need.

Experience: 2005-2006 IMI Vision UK and USA 12 months industrial placement with many roles including project management. 2007 MADE IN BRUNEL directory team, designing and co-creating this book.

T   +44 (0)7786 964014
E   jon@jgmarsh.com
W   www.jgmarsh.com

Lingo is a device that translates both audio and photographic content into any language, enabling the user to communicate with people whilst travelling the world. The brief for this project was to develop a product for the well-known Red Bull brand which would focus on vitalising the mind of the user whilst encouraging them to explore the world around them. Red Bull gives you wi-ings!

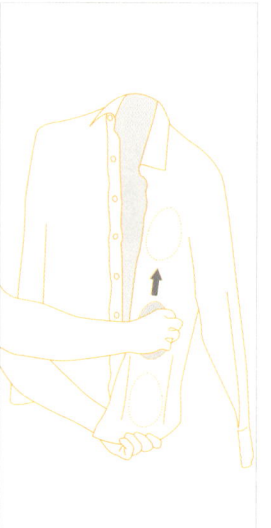

ion. is unlike any 'ironing' product that has previously existed, 'ion' removes the creases from clothes without the need for an ironing board. By using two hot plates which are magnetically coupled, heat and pressure are applied to both sides of the fabric resulting in a more efficient removal of creases.

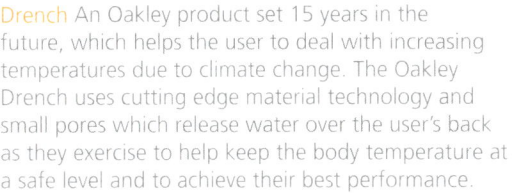

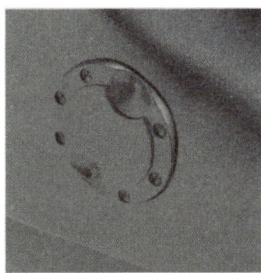

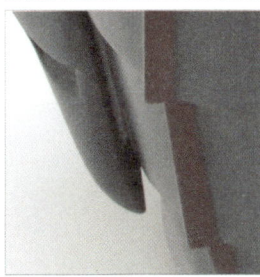

Drench An Oakley product set 15 years in the future, which helps the user to deal with increasing temperatures due to climate change. The Oakley Drench uses cutting edge material technology and small pores which release water over the user's back as they exercise to help keep the body temperature at a safe level and to achieve their best performance.

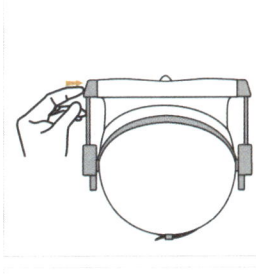

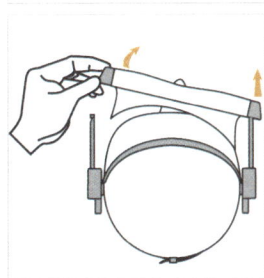

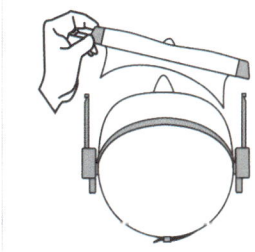

Dive mask attachment is designed specifically for hobby snorkellers to help with the removal and replacement while in the water. The product enables the user to remove and replace the mask with one single hand allowing them to clear it of any fog, while still being able to swim with the other hand.

STUART MASTERS

INDUSTRIAL DESIGN BSc

▬ Product Innovation ▬

Both the three years spent studying design at Brunel and my placement at No Climb Product Ltd. have allowed me to grow in confidence and expanded my knowledge of the processes and creativity needed by a successful designer. I have proficiency in many design software programs including Adobe Photoshop, Illustrator, Pro/ENGINEER, and SolidWorks as well as experience in both the Marketing and Technical departments at my placement company.

T   +44 (0)7793 028337
E   stuart_masters@hotmail.com

LIAM NEWLAND

INDUSTRIAL DESIGN BSc

Product Innovation

As a designer I love to be involved in every facet of design, I feel it is a designer's job to nurture a design until retail and I look forward to doing this in the future. I acquired training in CAD/CAM operation using programs such as Delcam and Rhinoceros and operating Datron CNC milling machines in my placement year at OTO 3D. The second half of my placement I put these skills to good use at Proctor and Gamble as one of their prototype manufacturers. I feel confident that the skills acquired on placement, added to those gained on the course, have enabled me to be a proficient and hard working designer with a love for challenges.

T   +44 (0)7738 477527
E   ljnewland@hotmail.co.uk

O₂ SUMMIT

ALTUS 1

Altus 1 a lightweight flat pack travel fin for scuba divers. The use of materials and design details enables the Altus 1 to be packed with ease whilst still performing well underwater. O2 SUMMIT is a high altitude-breathing mask for snowboarders and skiers which provides security and reassurance allowing them to reach new extremes. The Oakley product is targeted for a time 15 years in the future when global warming has diminished the lower snow peaks.

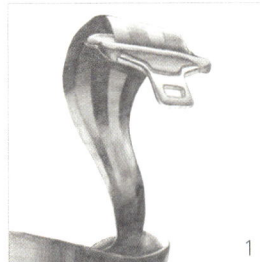

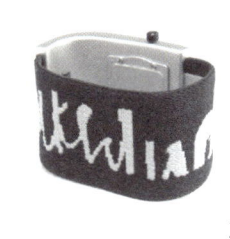

1

2

3

4

5

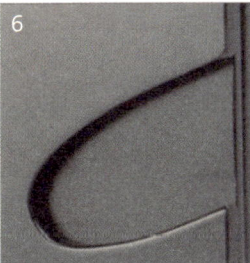

6

VIKAAS PATEL

INDUSTRIAL DESIGN BSc

As a designer, I like to make life as easy and efficient as possible. Understanding the problems facing society has helped me to develop a different approach of design. I seek to appreciate every product with minute detail and truly understand its purpose. The Brunel design process has helped me to establish my beliefs of what design is, by portraying them through my work and help me understand how a good innovation can achieve better design for people.

1 My interests are sketching and producing surreal drawings that use everyday ordinary products and place them into an exciting and unfamiliar context. 2 RUSH uses advanced technology to convert heart beat pulses into Quiksilver units. These are a conversion of the amount of adrenaline produced by the body from activities such as surfing. Part of the Quiksilver Sixth Sense range. 3 This self designed and written contextual essay looks at how the influence of sex appeal in product design has affected the consumer dislikes and likes of everyday products such as mobile phones, furniture and entertainment appliances. 4 – 6 'Utility Mat' is a car boot organiser. This enables more efficient and practical use of the space. The mat protects items during transit no matter how small they are.

T   +44 (0)7949 832101
E   vikaaspatel@hotmail.com

WILLIAM WARD

INDUSTRIAL DESIGN AND
TECHNOLOGY BA

— Product Innovation —

Studying at the University of
Technology, Sydney on exchange
has given me an invaluable
opportunity to experience a
different culture and to see
a new face of design. The
new skills that I have learnt in
Australia, coupled with three
years at Brunel, has enabled me
not only to incorporate suitable
aesthetics and graphical qualities
to a product, but to also take
into account the need to design
for manufacture and the various
considerations needed at each
stage of development.

T   +44 (0)7795 263379
E   will_ward_9@hotmail.com

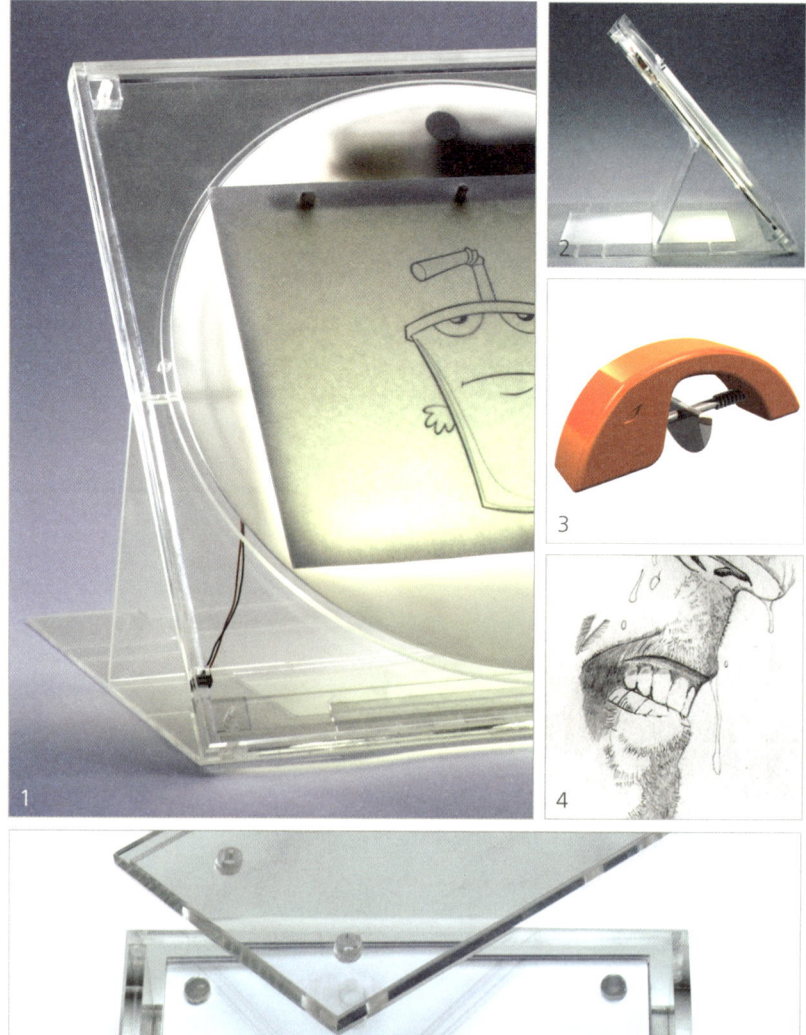

1, 2 This light box aims to better the performance of
existing products. Incorporates a battery and is able to
fit into a portfolio bag. 3 A concept herb dicer created
with the Alessi brand in mind. 4 Drawing is a main
hobby and pastime. 5 A personal branding exercise.

CHRIS WILSDON

PRODUCT DESIGN BSC

— Product Innovation —

I live for fresh air, photography and music but as a designer my main strength is a thorough understanding of software, particularly Adobe's Creative Suite and for 3D design, SolidWorks.

Please check out my website for further projects, photography and my full CV.

◄ Cross–section rendering of my final year major project, a Breath Trainer designed specifically for wind and brass musicians.

T   +44 (0)7729 604456
E   hello@chriswilsdon.com
W   www.chriswilsdon.com

INTERACTIVE MEDIA FUTURES

The digital age has wrought an explosion of
multimedia content, created and accessed
by anyone on the planet. The capability of
the internet as a connection to people on an
emotional level is only starting to be explored.
The potential for those taking our multimedia
to the next level is enormous.

MARIAN ABDULAHAD

MULTIMEDIA TECHNOLOGY
AND DESIGN BSc

As I have a strong background in art and an interest in computing, I endeavour to fuse the two disciplines to create one innovative design. My interests lie in creating multimedia artefacts that allow the user to experience something original.

I am also interested in exploring the impact multimedia has on the public as a whole. Ultimately, my enthusiasm for the subject drives me to pursue a career in this field.

T   +44 (0)7949 194135
E   marian_abs@hotmail.com
W   www.dive-inside.com

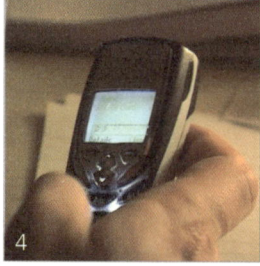

Ghost on Campus was the title of my final year project. Here I attempted to challenge the passiveness of simply watching a film by allowing the user to interact with it. Throughout the film the user is given a chance to change the course of events by selecting a different branch in the storyline. Opposite are stills from the film and DVD interface.

5 A QuickTime VR (a 360° view), created using VR Worx, based in Regent's Park Underground Station.
6 – 7 Examples of my photographic studio work and a sound project I created using Logic Pro and Final Cut Pro.

THOMAS ALEXANDER

MULTIMEDIA TECHNOLOGY
AND DESIGN BSc

— Interactive Media Futures ——

How can I explain myself and my design philosophy with mere words? I suppose this continual pursuit of innovation is the life of a designer. It is my belief that the day you believe your work to be perfect is the day you cease to be a designer. For me, design is a process of trying, learning and bettering oneself. And so my next step is to extend my passion for photo manipulation, illustrative design and mixed montage production into a profession.

This project explores the possibilities of realism in a fictitious world. By combining the narrative techniques of comics with photo realistic images, a popular children's book comes to life without the use of text. With a focus on manipulating and compositing images, this project aims to turn storybook illustrations into reality. Currently a semi-finalist at the Adobe Design Achievement Awards.

T   +44 (0)7730 594527
E   tom_a_gun@hotmail.com
W   www.designsandsolutions.com

23

ANDREW ARCHER

MULTIMEDIA TECHNOLOGY
AND DESIGN BSc

I am a Graduate in Multimedia
Technology and Design,
currently situated near Watford.
I play hockey for West Herts.
Hockey club, and am keen on
keeping fit and active. I've also
competed at County level for
athletics and table tennis. My
interests in multimedia include
both visual and audio. I play the
guitar and have a great interest
in computer gaming and CGI
film. I use all of these creative
influences to help me produce
a variety of New Media.

T   +44 (0)7738 276778
E   andy@flakx.com
W   www.Flakx.com/portfolio

My major project brief was to design a place where creative people,
who are interested in both visual and audio can come together. I
wanted to create a space where they could converse and contribute
to communal entities. The Flakx community can add to a Collaborative
Collage as well as Radio. The Collage allows for artists and designers
to add their images for display purposes, but also to create a larger
joint image. The Radio allows for music lovers and creators to add
their own music, or to publicise a band they like. The interface allows
both aspects of the creative community to see each other's work,
and add their own influence. This was an important aspect of my
project, because I didn't want one part of the community to be seen
as dominating over the other. The community can talk through a chat
interface and via forums. My project has utilised various technologies
including PHP, Adobe Flash, and Director Multiuser Server. Over the
course of my university career I've created a variety of Multimedia
Artefacts. These range from video to 2D/3D graphics.

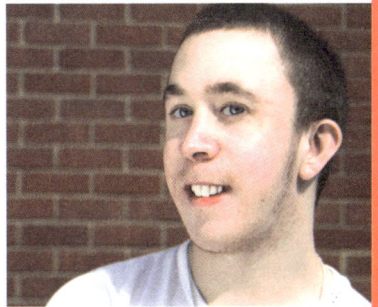

ADAM BAILEY

MULTIMEDIA TECHNOLOGY
AND DESIGN BSc

- Interactive Media Futures -

Having spent the last four years undertaking a degree, and an industrial work placement at SCT Ltd. during my third year of study, I am looking forward to being able to relate my skills and expertise to the practical world of multimedia design. My interests lie in game design, interactive interfaces, and 3D character modelling. I am passionate about creating 'that one idea nobody has thought of yet', and as such enjoy producing innovative and creative interactive media.

StreetVibe - "Created by Bournemouth, for Bournemouth" The latest in interactive data collaboration. StreetVibe is the online archiving of people's experiences and reviews of a location, which can in turn be explored by visitors wishing to follow in their footsteps. Tread dynamically through a resort at your own pace with help from a virtual friend.

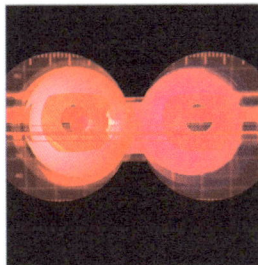

Reality Check - A fictional 3D movie trailer depicting various video game characters coming to life and breaking their programs - locked inside a computer world where the Anti Virus software controls all. Join their epic struggle as a hero goes about realising his destiny.

T   +44 (0)7709 103668
E   adz@adz-designs.com
W   www.adz-designs.com

TOM BREW

MULTIMEDIA TECHNOLOGY AND DESIGN BSc

Interactive Media Futures

When starting this course, they told us that we were all going to be Multimedia Architects – having creativity and the knowledge of how to use it, and they weren't wrong. My clean, minimalist design mentality was developed at Brunel and a year's placement at Webexpectations.com gave me the technical skills to make things happen. With a passion for HCI (Human-Computer Interaction) and appreciation of simple, good design, I hope to have an exciting multimedia future ahead of me.

T   +44 (0)7709 166937
    +44 (0)7624 472516
E   mail@tombrew.co.uk
W   www.tombrew.co.uk

e-Splart.com was my final year project. I explored the new way in which web users interact with websites and the new sense of community that brings. Using a virtual paint can, budding 'splartists' throw paint onto a canvas to create their own unique web-art.

At Brunel, I developed my 3D modelling and animation skills considerably. 1 The idea for my animated fly-through of a Lego City was born from my, probably unhealthy, fascination with Lego from a young age.

2 For my final 3D project, I decided to challenge myself and see if I could model a human body. Drawing inspiration from my passion for football and Arsenal FC in particular, I successfully modelled myself in various footballing poses.

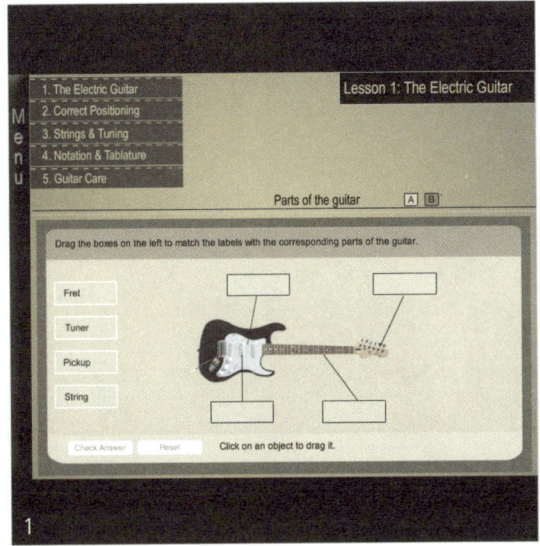

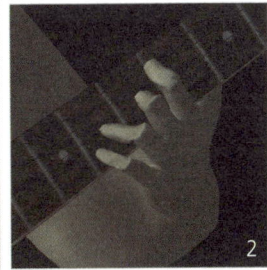

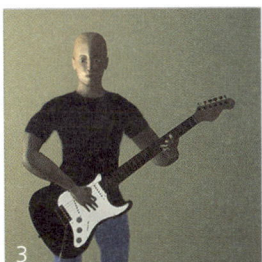

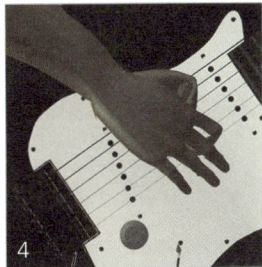

An interactive guide to the basics of playing the electric guitar, as demonstrated by a 3D character. The guide consists of a number of tutorials covering everything from the different parts of the guitar to reading tablature. **1** A drag and drop 'label the guitar' exercise from the tutorial. **2** Finger positioning. **3** Holding the guitar. **4** Plucking the strings.

5 An eyeball from a 3D animation about the consequences of bouncing around when you're an eye. **6** A frame from a 3D animation showing a shadow passing over a garden scene featuring a water fountain. **7** Characters from 'The Elementals', a 2D animation about a group of superheroes, each with the ability to control one of the elements. **8** A photorealistic 3D image of a living room.

NADIA HUSSAIN

MULTIMEDIA TECHNOLOGY
AND DESIGN BSc

— Interactive Media Futures —

As the illustrated project indicate, I have a keen interest in 3D work and music. I also enjoy illustration, website design, interactive media and filming. The diversity of the Brunel course has allowed the accumulation of skills in all of these aspects of the discipline. I have spent a year on an industrial placement as a Multimedia Assistant at the Wellcome Trust and I would like to further develop my career in 3D animation.

T   +44 (0)7950 223870
E   nadia@9musemedia.com
W   www.9musemedia.com

27

GIULIA MEEK

MULTIMEDIA TECHNOLOGY
AND DESIGN BSc

Over the last four years I've done
a lot at Brunel. I've had some
great experiences and made a lot
of good friends. Oh yes, I've also
studied a bit; I've learnt a great
deal about multimedia technology
and techniques and their various
aspects. My strengths have been
demonstrated in the areas of web
design, video editing, Adobe Flash
animation and sound production.
I spent my placement year
working as a web designer. The
last few years have been busy; I'm
going to miss Brunel, but I think
I'm ready for the next big step.

T   +44 (0)7815 163433
E   giulia@notjulia.com
W   www.notjulia.com

My final year project focused on the development of
an interactive guide to fan conventions. I thought it
would be an interesting idea to try and introduce this
community to an outsiders point of view. I created a
virtual tour of a con, created in Adobe Flash. Within
the environment the user can interact with the
characters, watch videos, play games and learn about
what goes on at conventions all over the world.

The brief was to edit together a video and create a
piece of music to accompany it. Having convention
footage at hand, I used it for the video. The result
was integrated into my final-year project.

ROB STEWART

MULTIMEDIA TECHNOLOGY
AND DESIGN BSc

World Cup Skiing Rebrand was aimed to design and implement a new brand for a skiing programme. I created a full graphics package from title sequence to leader boards. It required a range of skills in a number of design and video editing packages, including Viz Artist (a real time broadcast graphics system) and Microsoft Visual Basic.

I enjoy multimedia, learning about new technologies and developing creative designs that are both unique and innovative. I have always been fond of television and film, and have developed a background in television graphics, from my work placement at Wurmsers Ltd. My other love in life is sports, in particular badminton, rugby and football as I believe they are great ways to socialise and exercise. I aim to continue to be innovative as a multimedia developer throughout my professional career.

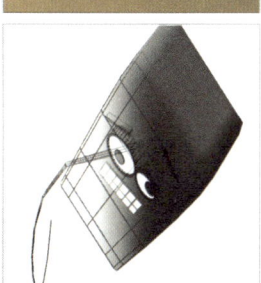

Working Late is a short story animated in Autodesk 3Ds Max, with additional compositing and editing in Adobe After Effects and Adobe Premiere respectively. It follows a simple narrative of a stapler working into the night who gets a staple stuck and, with some help, attempts to remove it.

T   +44 (0)7841 845619
E   rob@outlined.co.uk
W   www.outlined.co.uk

DAVID STRUGNELL

MULTIMEDIA TECHNOLOGY
AND DESIGN BSC

Interactive Media Futures —

When searching for a degree
I wanted to find a subject that
combined my fascination with
technology, computing and
design - Brunel's Multimedia
degree was ideal. Four years
on I have no regrets. After
acquiring a wide range of skills
in the first two years, I spent
the third year putting them into
practice and developing them
further during my placement at
IBM. After my year in industry
I felt ready to tackle my final
year, which included leading
the development of this year's
MADE IN BRUNEL website.

T   +44 (0)7971 574367
E   david@strugnell.com
W   www.strugnell.com

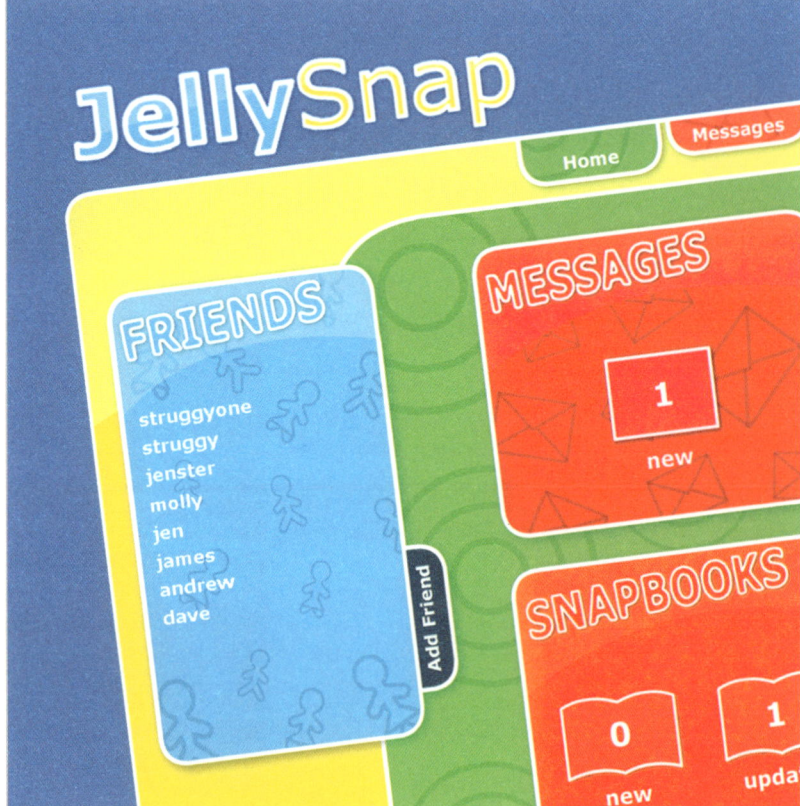

JellySnap was a project undertaken in my final year,
and was developed using Adobe Flash/ActionScript,
PHP and MySQL. It is a new socialising website
designed especially for children, providing them with
a safe and secure online environment where they can
share their photos and socialise with their friends.
To find out more please visit www.jellysnap.com.

Another skill I have developed during my time at Brunel
is 3D modelling and animation. The images above are
taken from 3 animations I created during my final year.
They were all produced using Autodesk 3ds Max with
additional editing and compositing done using Adobe
Premiere. To see the full animations along with the
rest of my portfolio please visit www.strugnell.com.

Q-Nav is a new and exciting mobile application tool primarily aimed at theme parks. Its designed to help people navigate a theme park more efficiently with access to ride queue times and news that's always up-to-date. Whatever the guest's location is in park, they can access live up-to-the-minute information on queue times on any ride in the park. They can also get the latest news on rides and shows.

MARK TUNNICLIFFE

MULTIMEDIA TECHNOLOGY
AND DESIGN BSC

■ Interactive Media Futures ■

I consider myself to be both a creative and a technical person, in creating my final year project 'Q-Nav' I had to create a project that would show my full potential in both areas, to ensure I could get the best possible mark. Since joining my course at Brunel it has constantly made me think in new and creative ways, which I believe is shown in my projects.

Q-Nav is designed to run on mobile phones, it has a simple yet creative interface to allow park guests to access ride information. All the data on ride queue times and news are stored externally from the phone, this allows the application to be flexible and very dynamic as it can be constantly updated.

T +44 (0)7739172498
E mark.tunnicliffe@gmail.com
W www.tunnie.co.uk

AMARDEEP AND GURMIT SHAKHON

MULTIMEDIA TECHNOLOGY
AND DESIGN BSC

Interactive Media Futures

Urban Spaces was a project that justifiably required the efforts of two multimedia practitioners due to its scale and complexity. Amardeep and Gurmit Singh Shakhon are the proud owners, their history together working on multiple projects as freelancers and being best friends allowed them to pursue this journey together, this has by far been the most challenging project tackled.

"We set our standards extremely high, as we wish to continually improve our skills and those whom we work with."

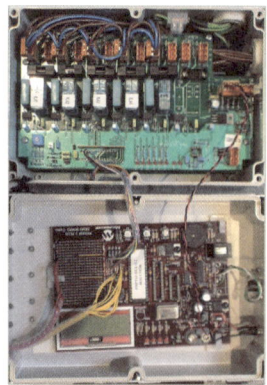

Urban Spaces was created on the foundations of HCI practitioner's findings and conclusions. To extract data on participant's ability to navigate through a controlled environment of altering light and space. The artefact uses doors, florescent tube lights, pressure pads and visual display units to create a unique abstract environment that is highly interactive and provides an innovative and unique user experience.

The aim of this project is to aid understanding into human navigation and how humans base their decision-making abilities within both physical and virtual environments. Through the use of various pervasive input devices and C programming alongside a Java interface we are able to record the human navigation process without influence or manipulation to the outcome.

T   +44 (0)7985 709962
    +44 (0)7985 709924
E   info@mapleonline.co.uk
W   www.mapleonline.co.uk

DESIGN AND SOCIETY

Modern society is influenced by a large number of factors; and many of these, be they political, environmental or economic, are intrinsically linked to design. This gives the next generation of designers both powerful opportunities and critical responsibilities.

JADE HUTCHINSON

INDUSTRIAL DESIGN BSc

— Design and Society —

The value of design is now being measured by its impact on business. I am a strategic thinker, who is ready to assist companies prepare for the future. I have gained valuable work experience at Integrity Design Management, working with brands such as Marks and Spencer, Volkswagen and Virgin.

T +44 (0)7851 751942
E jade.hutchinson@gmail.com
W www.coroflot.com/
  jadehutchinson

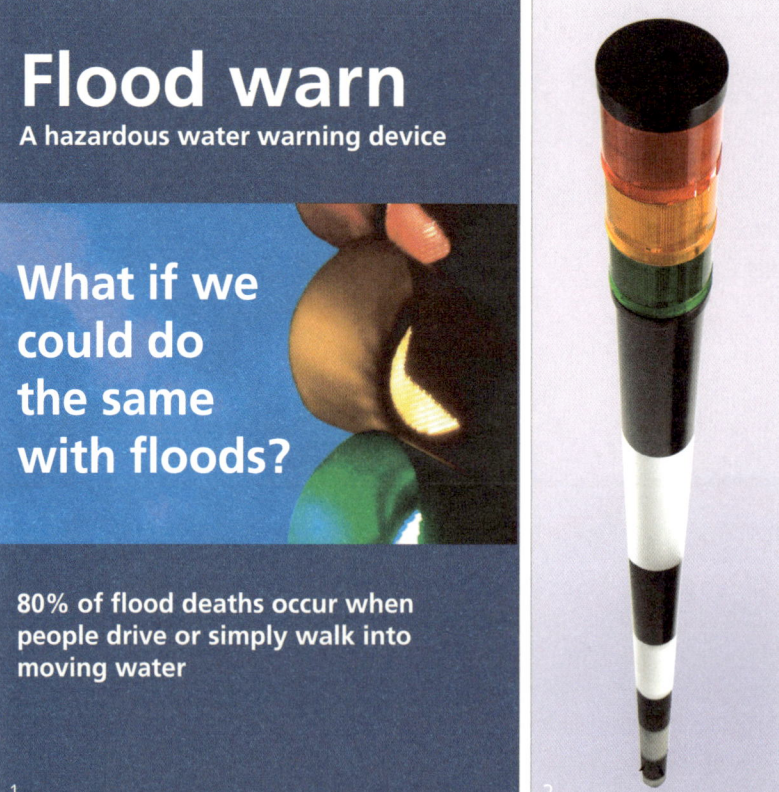

Flood warn
A hazardous water warning device

What if we could do the same with floods?

80% of flood deaths occur when people drive or simply walk into moving water

1

2

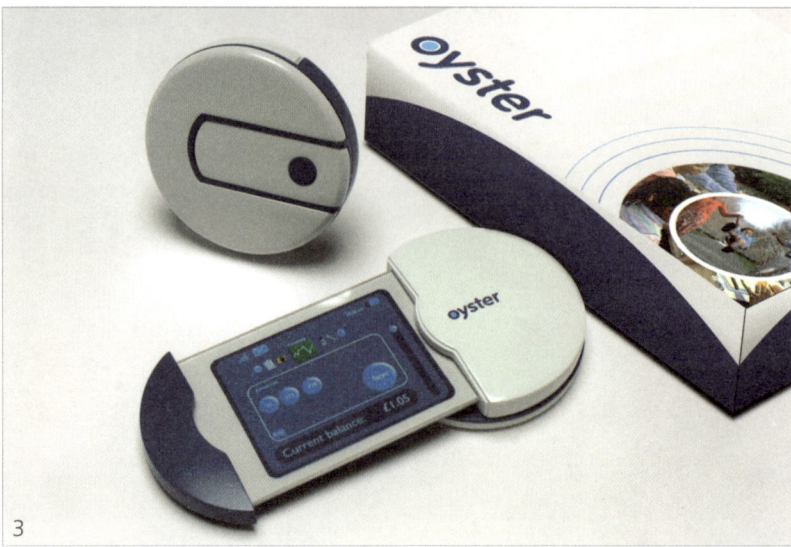

3

1, 2 Flood warn is a hazardous water warning device that communicates the treat level of flood waters using our understanding of colour. **3** Oyster Family - Stay in touch with family and friends, control your kids spending and store membership information safely, all in the palm of your hand.

1 Project 'De-Mock'racy is the name of my final year Project and was created using Adobe Photoshop and Macromedia Dreamweaver. In the last two General Elections, the number of 18 to 24 year olds who voted has been worryingly decreasing. I created a website that would appeal to that specific target audience in order to engage them within politics.

A module that I found both challenging and rewarding in my final year was 3D animation. I had always wanted to tackle this subject and Brunel gave me the teaching and equipment necessary to do so.
2, 3 Some of my 3D work undertaken this year.
4 An example of my graphic design work, achieved by using Adobe Illustrator and Photoshop.

GHAALIB KHAN

MULTIMEDIA TECHNOLOGY
AND DESIGN BSc

— Design and Society —

Graphic Design has been a hobby of mine for many years and when coming to university I wanted the opportunity to turn my passion into a career. Multimedia Technology and Design at Brunel turned out to be the perfect course for me as it further developed my graphical skills and also introduced me to areas such as 3D, digital sound and video. The skills and confidence gained from my time at university has enabled me to conduct my own freelance work. This has given me valuable experience of working with clients to achieve common goals.

T   +44 (0)7791 046769
E   gkhan@ghaalibhhan.co.uk
W   www.ghaalibkhan.co.uk

CHAO-CHI LIN

MULTIMEDIA TECHNOLOGY
AND DESIGN BSc

Design and Society

I have heartily committed myself to design throughout my life. I try to enable and encourage sophisticated thinking about technology in relation to culture and perception. My artwork relates to that part of my brain that contradicts articulation. Being a fan of simple lines and colours, I try to communicate by such elements and bring out my view of life.

My project is aimed at the Taiwanese market and has been developed using 3 characters. The connection between people's living and products, along with interactive design is brought together on a website. This provides a social exchange of knowledge and experience, introducing the concept of the nostalgic products via digital media and internet, to communicate our common daily lives and culture.

T +44 (0)7921 145787
E mantoboo@googlemail.com

The Iron tower is away from 1000000 miles

My project is centred around a completely
original 3D design which incorporates
another personal passion, architecture.

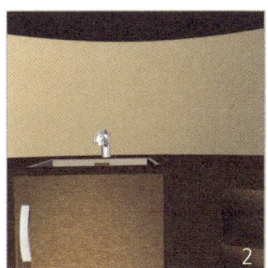  

1 A render of my completed building
2 An interior render 3 A photorealistic scene
4 A Lego Technic Mini Loader.

LUISA
SOUTHGATE

MULTIMEDIA TECHNOLOGY
AND DESIGN BSc

- Design and Society -

For these last four years I've
been exposed to every media
imaginable, including sound, 3D,
web design, and film. Uniquely
I've not only been taught the
technology behind each media,
but how to use them, both
together and apart, effectively. I
have also had the opportunity to
put these new skills into practise
by doing a work placement for
my third year, with a company
that focuses upon the post-
production side of the film
business. Learning at Brunel has
been an honour, and I can't wait
for the next challenge.

T +44 (0)7814 259156
E casinofall@hotmail.com

JAY SUTHAR

INDUSTRIAL DESIGN BSc

— Design and Society —

Experience: 13 months design placement with Eleksen, a world leader in smart fabrics technology, where I primarily acted as a web and graphics designer, and later involved with product design on projects for Nokia, Boeing and Microsoft. Projects included collaborating closely with marketing and product managers as well as being involved with engineering.

This professional development, coupled with the schooling from the Brunel Design course has been fundamental to my progression as a designer possessing a mixture of creative and technical dexterity necessary for industry. I believe the understanding of technology, style, function, commercial practicality and the requirements of the end user are crucial to achieving a good design solution.

T   +44 (0)7779 296659
E   sutharjay@hotmail.com
W   www.coroflot.com/sutharjay

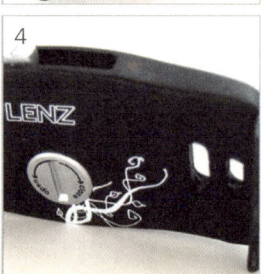

LENZ a head mounted waterproof camera for capturing surfing moments. The camera itself is compact, lightweight and ergonomic in its design – a quick slide of a button can allow you to quickly start filming footage. The Neoprene head strap ensures that the product fits securely and comfortably onto the head. Part of the Quiksilver Sixth Sense product range.

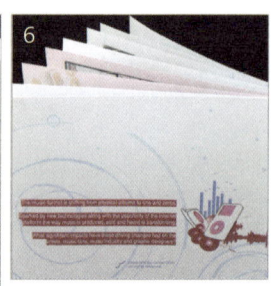

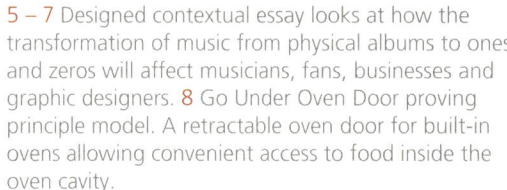

5 – 7 Designed contextual essay looks at how the transformation of music from physical albums to ones and zeros will affect musicians, fans, businesses and graphic designers. 8 Go Under Oven Door proving principle model. A retractable oven door for built-in ovens allowing convenient access to food inside the oven cavity.

CLAIRE WALKER

MULTIMEDIA TECHNOLOGY
AND DESIGN BSc

— Design and Society ————

Originally from Norfolk and still supporting Canaries FC, showing loyalty before glory if nothing else. I am now London-based and exploiting all the city has to offer with regards to life, culture and experience. My interests lie in photography, stage and TV environments, utilising these mediums to generate intrigue and to captivate an audience's imagination and curiosity. With a diligent and flexible nature I am always up for a challenge and willing to have a go at almost everything (once).

A documentary on... is a short film about immigration in the UK today, it takes views and opinions of the general public, and then looks specifically at two connected yet distinct groups of immigrants. Film, edit and aesthetic styles vary throughout the piece giving movement and interest whilst still maintaining a consistent experience.

T   +44 (0)7919 592132
E   mzcwalker@hotmail.com

COMMUNICATION FUTURES

One basic human requirement is that of contact
with other people. The ways in which we can
now do so are continually expanding in number,
and in this sector new ideas and technologies
are emerging that seek to promote communities
that unite people across the globe.

SEBASTIAN FLIPPENCE

MULTIMEDIA TECHNOLOGY AND DESIGN BSc

► Communications Futures ◄

Studying at Brunel has allowed me to expand on my technical and design skills and as the final year project by nature has been left vague and ambiguous, which I believe has left the door open for any type of project that falls under the umbrella of multimedia design. This openness allowed me to choose to extend my knowledge on technologies and designs that have been used to create a socially integrated website.

T   +44 (0)7749 192997
E   seb@sebflipper.co.uk
W   www.sebflipper.co.uk

DuoMesh A new and exciting way to organise, share, explore and keep up to date on topics and media that are important to you, whether you're on the web, using a mobile phone or alternative mobile device. It is designed with synchronisation in mind, keeping you up to date while you're on the move, at work, or at home. It helps you find people with similar interests and ideas, and also helps you find related topics that interest you.

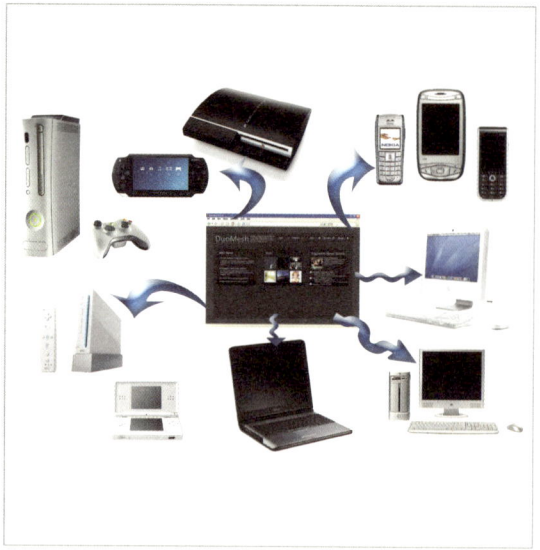

DuoMesh is a dynamic, intuitive website for aggregating and distributing user submitted content found on the internet. It supports XML/RSS content known as feeds and allows users to see which are the most popular, as well as being able to browse through catalogues of topics known as tags. Users can create a personalised homepage whilst also keeping track of other users by adding them as friends. The feeds and homepage can also be redistributed back into RSS to allow users to reuse the content in different contexts.

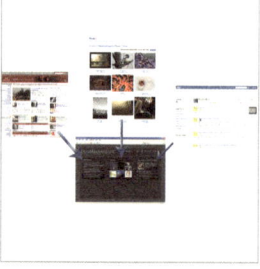

JESSICA NG

INDUSTRIAL DESIGN
AND TECHNOLOGY BA

— Communications Futures —

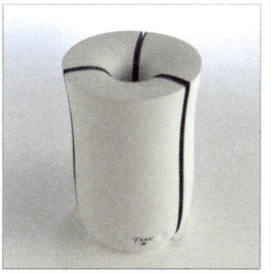  

Reminis inspired by Dove's 'campaign for real beauty', enhances the mood of its users thought the application of visual stimulation. As part of the proposed 'feeling good' range, this modular, 3D holographic projector aims to provide the male market with a range of products that offer a sense of reassurance, security and support.

An enthusiastic designer, I have been lucky enough to have experienced all areas of design. Having seen many projects from the initial concept development stage thought to realisation.

I have developed a well rounded understanding of the design process. My time spent in China also provided me with invaluable life experiences and a more culturally diverse view of industry. My main areas of interests and my strengths centre around the aesthetic elements of design.

EMU a set of magnetic alphabet lights based on the idea of fridge magnets have been designed to encourage young children to explore the English language through the use of play. Large, tactile and colourful, the soft outer shell invites interaction. Based on a simple LED circuit, the lights are both efficient and fun.

T   +44 (0)7969 963768
E   jessng85@hotmail.com

MATTHEW OAKES

MULTIMEDIA TECHNOLOGY
AND DESIGN BSc

— Communications Futures —

My experiences over the last
four years at university have
helped me develop my skills as
a designer while fuelling my
passion for creative multimedia.
My industry year helped me focus
on the aspects of design I really
enjoyed, as well as introducing
me to the world of work. After
accomplishing so much in such
a short amount of time, I can
only wait in anticipation at
what the future may hold.

T +44 (0)7797 784100
E matt@newmediaexperience.com
W www.newmediaexperience.com

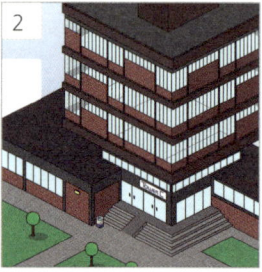

1 – 4 A fun and attractive website using PHP, Adobe
Flash and MySQL, to aid in the communication process
between the university and students within the
multimedia team.

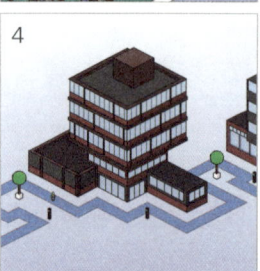

5 – 8 These pieces were created for my 3D assignments
during my final year. The main photo shows a soldier of
the future in full body armour, created in AutoDesk 3ds
Max and Bryce.

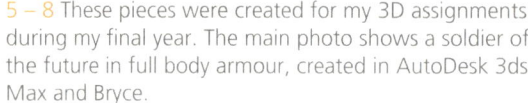

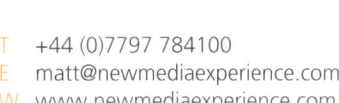

YONE SANTANA

MULTIMEDIA TECHNOLOGY
AND DESIGN BSc

I've been drawing since I was young and I've had the fortune of working as a Concept Artist, Animator and Web Designer. I was born in London but grew up in Spain and studied Architecture in Madrid as well as Illustration. I came back to the UK to study at Brunel and to continue my career.

◀ Character animation is an ongoing process from the initial conception of an idea, through sketches to the final rendered 3D animation. Multimedia is a great field to work in as it allows one to bring together different techniques from traditional art and blend them with technology to create something unique.

T   +44 (0)7913 483974
E   yone.santana@tiscali.co.uk
W   www.yonesantana.com

CHARLIE SPENCER

PRODUCT DESIGN BSc

Design, from graphical to product solutions, has always been a personal passion, as it allows us to have a highly visible impact on the world around us. Being part of Brunel Design in particular has developed my passion for creating exciting, innovative and dynamic projects. The broad range of modules tackled, and a year's work experience with Schefenacker Vision Systems, have provided me with a diverse introduction to the world of design.

Snow Glove Communicator System [SGCS]

Integrated 2 way radio/winter sports glove with a revolutionary interface design. Using specifically engineered electrotextile materials to measure finger positions (in terms of flexion or extension) the glove can interpret your hand gestures and deliver appropriate signals to embedded electronic equipment. Careful design inhibits transient activation, so only when desired are actions carried out by the onboard microcontroller. The use of stereotypical gestures as activation methods for electronic functions means controlling the equipment is quickly learnt and easily remembered, to the point that the interface becomes second nature to use. The SGCS is designed to act as a more efficient method of communication on the slopes market research conducted showed consumer displeasure at current solutions), however the system technology can be utilised in various fields.

T +44 (0)7738 430811
E charlie@breakthecircle.info
W www.breakthecircle.info

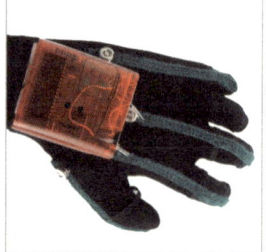

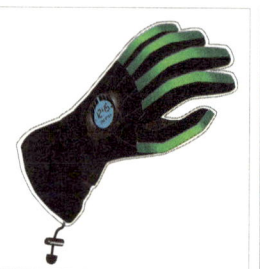

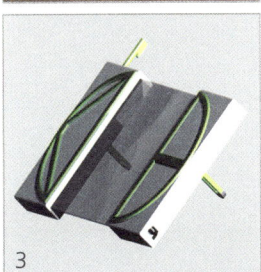

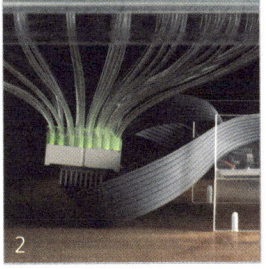

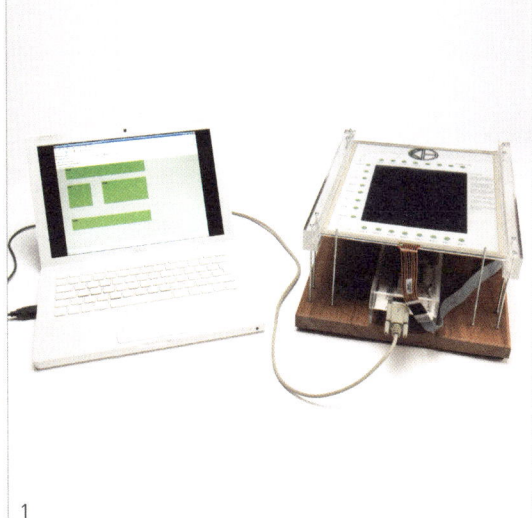

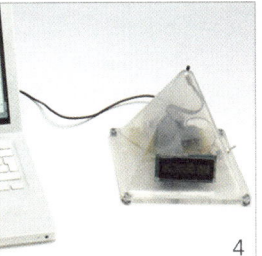

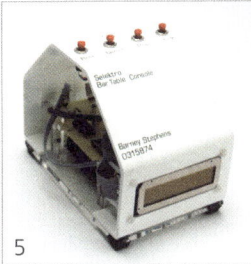

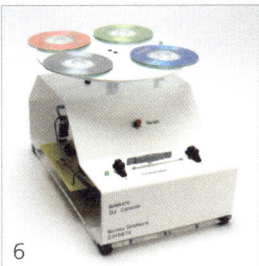

BARNEY STEPHENS

INDUSTRIAL DESIGN
AND TECHNOLOGY BA

- Communications Futures -

Experience: 2005 – 2006
ChameleonNet, Junior
Web Designer.

Skills: XHTML, CSS and
FTP; Adobe Photoshop,
Illustrator, InDesign, Flash and
Dreamweaver; SolidWorks,
AutoCAD and MPLab.

Interests: Web design, DJing,
music production and travelling.

1, 2 The Karma Board is a tangible system for designing and coding website layouts or wireframes.
3 A concept rendering of The Karma Board.
4 ThermoMIDI is a temperature controlled MIDI instrument. 5, 6 Selektro is an automatic voting and selection system for use in nightclubs. 7, 8 Personal logo and portfolio design.

T   +44 (0)7795 833803
E   barney.stephens@gmail.com
W   www.barneystephens.com

INNOVATION FOR AN ACTIVE LIFESTYLE

Outdoor pursuits and extreme sports form part of a global industry that is continually developing to meet the needs of athletes and adventurers alike. As a trendsetting market in many ways, innovation is an important characteristic for any successful design in this field.

MATTHEW BARNETT

INDUSTRIAL DESIGN BSc

- Innovation for an Active Lifestyle -

I will be the first to admit a rather obscure imagination is the driving force behind my creativity. Working with design-led J-me, and expanding my horizons studying architecture and furniture design at UTS, Sydney, has led to a design thinking greatly influenced by our emotive interactions, and the implementation of these ideas to marketable products. Beyond design I release my energy via my photography, and worked with the MADE IN BRUNEL team to produce the portraits within this directory.

T   +44 (0)7771 640140
E   dt03mbb@hotmail.co.uk

1

2

3

1 Does it only snow in winter? Snowdown creates snowballs to provide refreshing fun and promote activity within children all year around. 2 Stroke – an interactive gel light. 3 Marmite Lasertag pieces morph to a weapon when an opponent approaches, only the quickest survive. Aimed to upset the mundane, it is an emotive product for an emotive brand.

1 The Automatic Swim Lap Counter - This fully functional solution detects a specific individual. The compact, battery operated product is totally waterproof, giving audio and visual feedback on the number of laps completed, regardless of the presence of other swimmers in the pool. A photo sensor unit at the poolside detects a reflective headband, reliably counting the laps so that the swimmer doesn't have to.

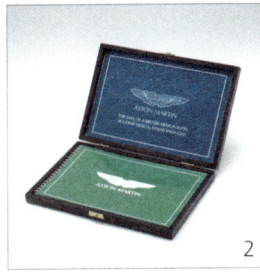

2 A contextual essay on the future of Aston Martin who were recently bought by ProDrive.
3, 4 The Mr Kipling Brand Analysis Report and the Mr Kipling Induction Program - A futuristic solution for sharing information with new employees.

JULIAN CHARITY

INDUSTRIAL DESIGN
AND TECHNOLOGY BA

Innovation for an Active Lifestyle

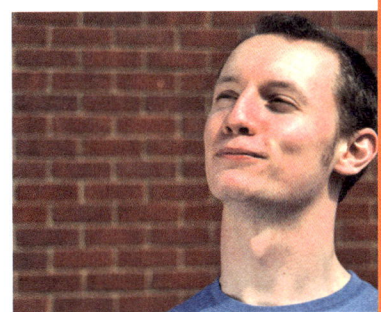

Experience working at automotive company Schefenacker Vision Systems has increased my knowledge of design for manufacture and assembly. My year with Schefenacker allowed me to work on projects for prestigious companies such as Toyota and Land Rover. I am a keen athlete and sportsman, I also enjoy flying in my spare time.

T   +44 (0)1476 563896
E   juliancharity@hotmail.com

JEREMY CROUCH

INDUSTRIAL DESIGN AND TECHNOLOGY BA

Innovation for an Active Lifestyle

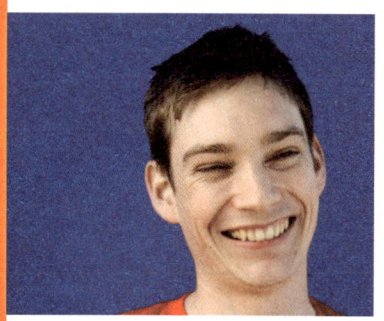

Learning at Brunel has given me a range of skills that allow the realisation of products through creative thinking and the application of technical functionality. I am also an outdoor enthusiast with a passion for mountain biking and associated technology. I look forward to a career in design that will complement this.

T   +44 (0)7984 585414
E   jezcrouch@hotmail.com

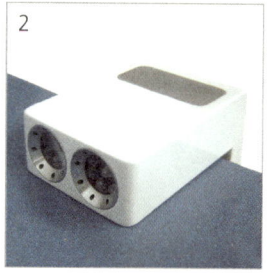

1 Red Bull Water Wing is a self-propelled float for children, giving them the confidence to swim.
2 Bush Baby is a touch-activated desk light that magnetically clips to a power rail on the desk's edge.
3 Mechanical Clock is a motor driven mechanical clock that keeps itself in time using a PIC microcontroller.

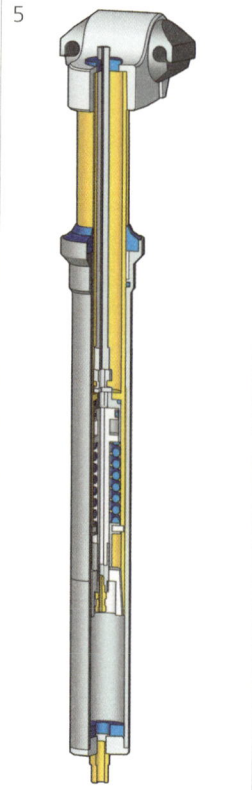

4 The Adjust-On-The-Fly Seatpost is a lightweight seatpost that allows the rider to change their seat height by up to150mm whilst still moving, using a hydraulically released, spring-loaded taper clamp.
5 Adjust-on-the-Fly Seatpost CAD Cut-Away.
6 Adjust-on-the-Fly Seatpost Handlebar Lever.

JEFFREY KNAPMAN

INDUSTRIAL DESIGN
AND TECHNOLOGY BA

- Innovation for an Active Lifestyle -

A placement year with
Alvan Blanch Dev. Co. Ltd.
allowed me to build on the skills
learnt at Brunel and encouraged a
'hands-on' approach to problem
solving. Large scale agro-industrial
three-phase machinery and
PLCs provided a fascinating
contrast to blue foam, product
styling and microcontrollers.

◀ Sea View wireless eyewear for
marine applications enabling the
hands-free viewing of an array of
navigational data, anywhere on
the vessel. 1 – 3 Mr Kipling 'Little
Helpers' – a shopping trolley
mounted electronic assistant
manages the parents' shopping
list whilst simultaneously
communicating with the kids'
handheld modules, instructing
on items to collect. Designed
to encourage family bonding.

T   +44 (0)7920 422394
E   jeffknapman@hotmail.com

WILL POSTLE

INDUSTRIAL DESIGN
AND TECHNOLOGY BA

- Innovation for an Active Lifestyle -

My enthusiasm towards design is a clear reflection of my time at Brunel University and San Francisco State University. Studying abroad for a year and having travelled extensively has brought about a cultural influence towards my thinking. I have an adventurous, artistic nature, which pushes my creativity with every new challenge I face.

Software Skills: Mac and Windows, Pro/ENGINEER, SolidWorks, Autodesk AliasStudio, Adobe Photoshop, Adobe Illustrator, Adobe InDesign, Adobe Flash, Autodesk 3ds Max, MPLAB.

Personal Skills: Rapid visualisation, model making, workshop skills including plastic, metal and wood.

Interests: Scuba diving, snowboarding, sailing, travelling and Burning Man

T   +44 (0)7786 116684
E   willpostle@gmail.com
W   www.willpostle.com

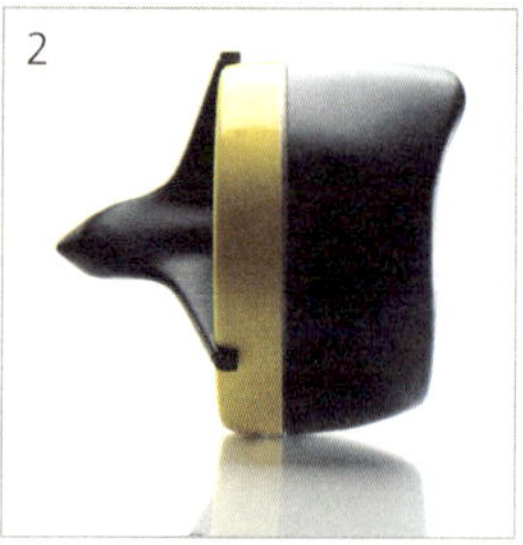

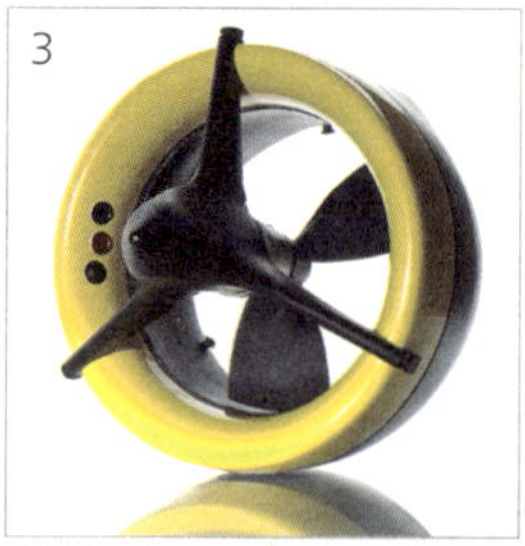

1 Vapour enhances the underwater experience for scuba divers. Utilising the high energy densities of lithium-ion cells, a unique shroud design and brushless DC motor technology, Vapour can tow a fully equipped diver at 2.24 m/s. 2, 3 A water tested working prototype of Vapour. 4 A book presenting a debate on the life of El 'Che' Guevara. 5, 6, 8 Lovers and Haters have become Heroes and Villains for Marmite's 110th Anniversary 'Marsquerade' Ball. 7 Standing at 1.5m tall, Exstillo has been developed based on the effects of human emotion to moving light.

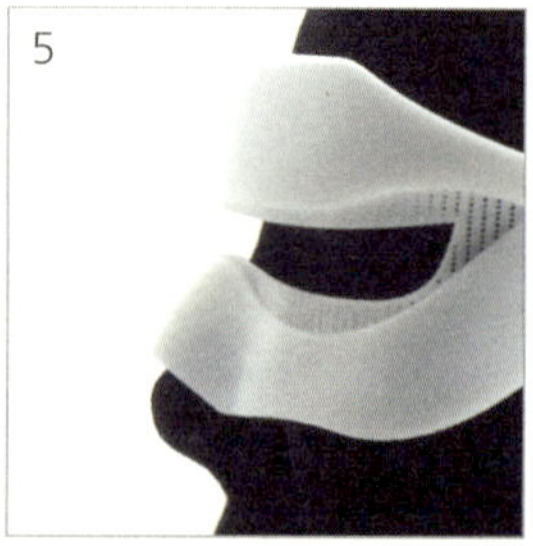

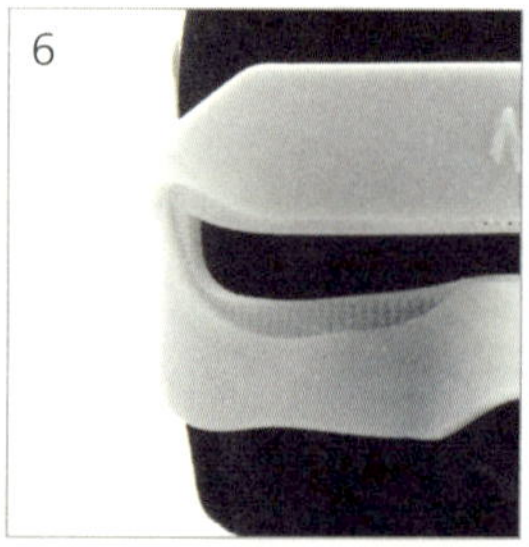

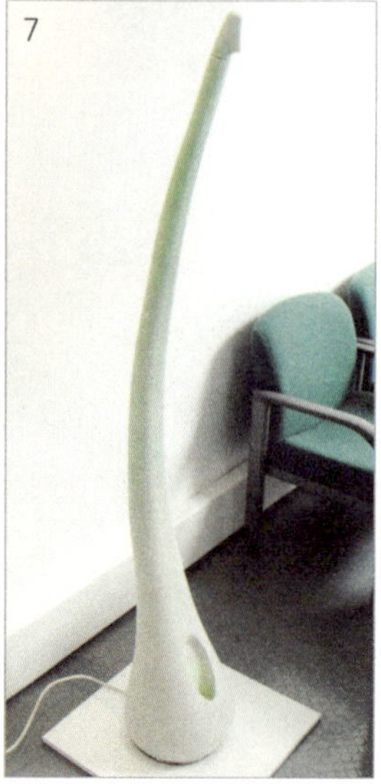

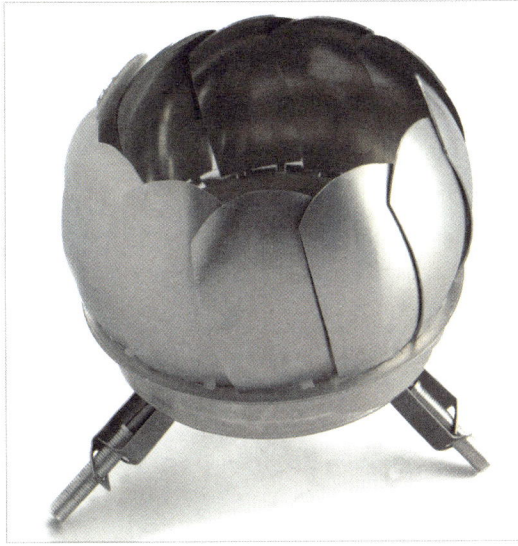

SOPHIA RICHARDS

INDUSTRIAL DESIGN BSc

‑ Innovation for an Active Lifestyle ‑

The stove stabiliser can level a camping stove on any terrain, with its easy to adjust tripod system and a spirit level which allows the user to ensure it is perfectly level every time. The leaves act as a windshield and the whole system folds away neatly, containing the stove for compact storage.

During my time at Brunel I have faced a wide range of challenges, from keeping to strict deadlines to racing yachts at student nationals. Each of these challenges has made me stronger as a person and furthered my understanding of design. As a designer, I want to use that understanding to encourage others to get involved with activities which are close to my heart. My placement has shown me what it is like working in a busy multinational environment and has developed my graphic design to a professional standard.

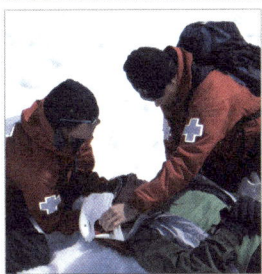

Always confident, protected, reassured. The Always first aid kit aims to encourage women to take part in outdoor sports such as mountaineering and white water kayaking. The first aid kit contains specially modified bandages that you can use one handed on yourself, as well as a GPS enabled personal locator beacon, which once activated can locate you in just 5 minutes.

T   +44 (0)7973 159784
E   sophia_richards@hotmail.com

JAMES STREATFEILD

INDUSTRIAL DESIGN
AND TECHNOLOGY BA

– Innovation for an Active Lifestyle –

Experience: 13 months placement with metal fabrication company, Stonebank Ironcraft Ltd., which is involved with designing and manufacturing bespoke metal pieces.

Responsibilities: Site Construction Manager.

Skills: Welding, spraying, metal fabrication, Adobe Illustrator, Photoshop, Microchip PIC programming, modelmaking, precise design and manufacture.

T   +44 (0)7841 835833
E   jstreatfeild@gmail.com

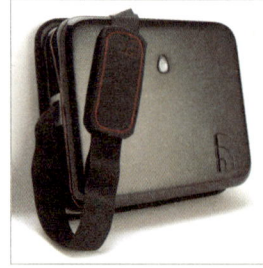

Kite Camera attaches to large sport kites and captures images of the kite operator. The camera connects via wireless technology to a viewer/recorder on the ground. A rotating weighted pendulum system ensures the image is constantly vertical even when the kite manoeuvres through turns and rotates 360°.
North Face Anti-Snatch Laptop Bag ensures safety of contents by using fingerprint recognition technology.

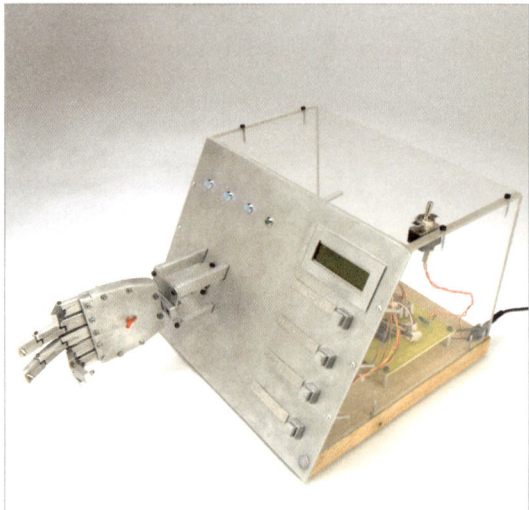

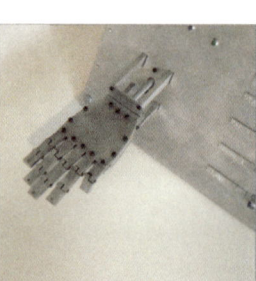

P,S,S An interactive paper, scissors, stone game. A PIC-controlled mechanical hand selects its own random sequences and displays its decision. An answer entry system combined with a LCD screen displays the results.

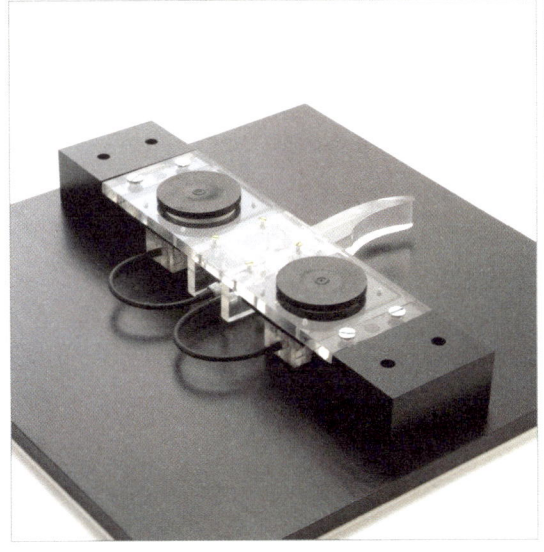

CHRISTOPHER WALKER

INDUSTRIAL DESIGN
AND TECHNOLOGY BA

Innovation for an Active Lifestyle

Experience: 9 months work placement with Pendennis Worldclass Superyachts.

Skills: AutoCAD, Pro/ENGINEER, Adobe Illustrator, Photoshop and InDesign.

Revolutionary Water Ski Binding The design utilises suction cups to enable rapid connection between the boot and the ski. The design provides more refined feedback to the skier and has been developed to be adjustable for different sizes and weights of skiers for improved safety.

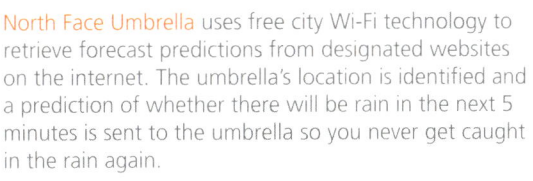

North Face Umbrella uses free city Wi-Fi technology to retrieve forecast predictions from designated websites on the internet. The umbrella's location is identified and a prediction of whether there will be rain in the next 5 minutes is sent to the umbrella so you never get caught in the rain again.

T   +44 (0)7977 427540
E   chris.walker84@gmail.com

GREG WASS

MECHANICAL ENGINEERING
AND DESIGN MEng

The MEAD course has given
me a firm grounding in both
Mechanical Engineering
and Design, allowing me to
apply technical knowledge
throughout the design process
in order to develop products
that are creative, functional,
and optimised for their purpose.
During my industrial placement
I worked for Cardinal Health
developing medical devices
such as infusion pumps and
disposable sets. I thoroughly
enjoyed this work and took the
lead in some very interesting
projects from the outset.

T   +44 (0)7867 503677
E   greg_wass@hotmail.com

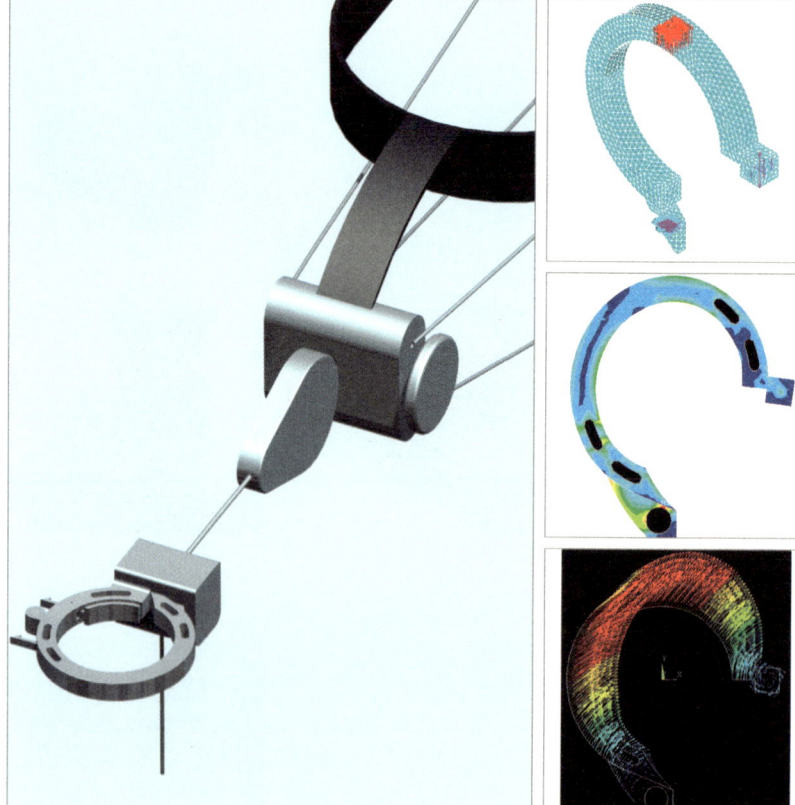

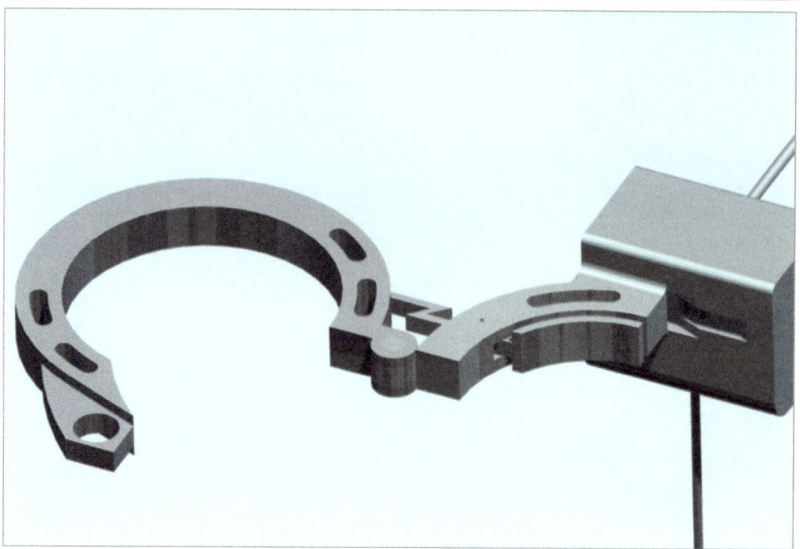

Snowboarders Lift Aid allows riders to comfortably use drag lifts
by transferring the force from the lift to their trailing hip. The
attachment clamp is on a recoiling mechanism, and includes an
innovative locking mechanism to ensure reliable attachment.
There is also a secondary quick release mechanism for additional
safety. All mechanical components were optimised in terms of
their strength to weight characteristics in ANSYS, and the unique
locking mechanism was proven using MSC Adams.

ELECTRONICS FOR TOMORROW

In an age of technological convergence,
electronics are the driving force behind every
modern appliance and communication system
designed to make our lives manageable.
Technical innovators are leading the way
towards advanced modes of living.

RAJIV BOSE

ELECTRONIC AND ELECTRICAL
ENGINEERING BEng

Electronics for Tomorrow

This final year project has been
ambitious; demonstrating
feasibility has always been
the target of this conceptual
project. The learning process
required to complete an end-to-
end working model has been a
challenge and a pleasure. Success
in this project has secured me a
collaborative PhD research post
between Brunel University and
Bradford University which is due
to commence in October 2007.
This project as it stands only
marks the beginning. Thanks to
the staff of the ISTA research
group of Brunel University for
their continued support.

T   +44 (0)7816 147349
E   rajiv_bose@hotmail.co.uk

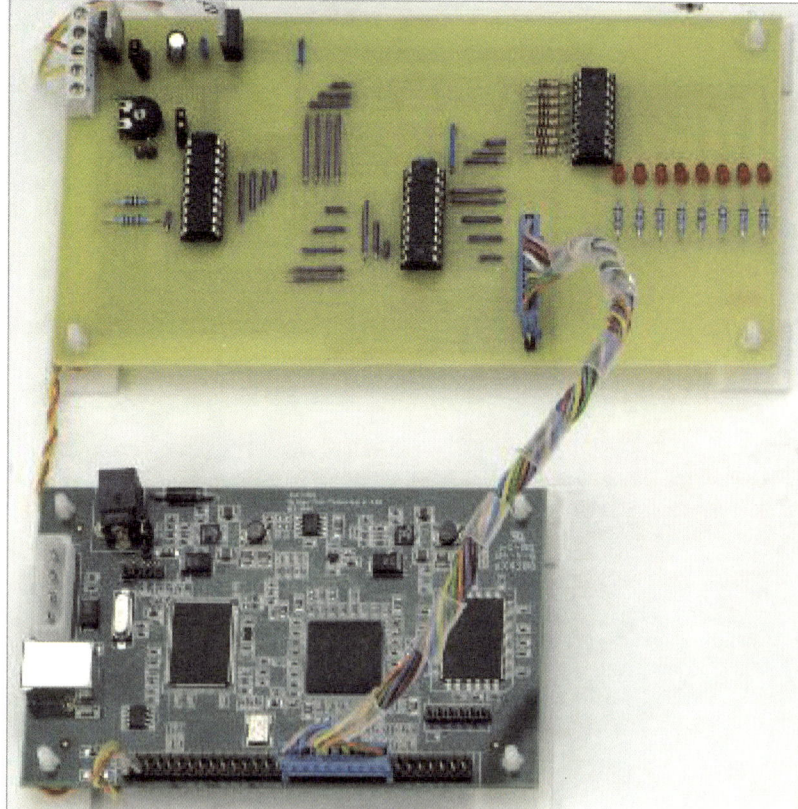

The above photograph shows an image of a high
end ADC circuit (Analog Devices Inc.) working in
conjunction with an FPGA development board (Orange
Tree Technologies Ltd.). This system configuration
marks the next generation of scientific instrumentation
for electromagnetic x-ray wave measurement.

The ADC chip ensures quantisation of a CCD's
output, while the processing capacity of Xilinx
Spartan-3 FPGA means there is no loss on CCD
resolution. Using a PDA as a user interface the
results are displayed in a concise graphical output.
Working as a stand alone system the potential for
mobile instrumentation becomes real, and precision
measurement is taken out of the lab and into the field.

Real-time streaming of audiovisual content over the Internet is emerging as an important technology area. Due to the wide variation of available bandwidth over Internet sessions, there is a need for scalable video coding methods and flexible streaming approaches that are capable of adapting to changing network conditions in real time.

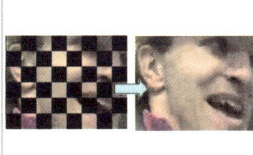

Using Joint Scalable Video Model (JSVM), Low-Density Parity-Check code (LDPC) software to implement Error Resilience using Forward Error Correction (FEC) in scalable H.264 for streaming video over an uncontrolled packet-based network channel. The scalability can be based on the video resolution or the frame rate depending on the application. Error in video transmitted, effects the quality, hence the main aims and objectives were to resolve this issue.

AMIT GANATRA

COMPUTER SYSTEMS AND ENGINEERING BEng

■ Electronics for Tomorrow ■

Work experience: technical support, system administration, Web development, and Consultancy. I am also a member of the IET. I would like to accredit my success and achievement in the completion of this project to Brunel University, The School of Engineering and Design, and my project supervisor: Dr. Jonathan Loo. Additionally, I would like to take this opportunity to thank my father Ashokkumar O. Ganatra, my uncle/sponsor Late Rajendrakumar O. Ganatra, my entire family and friends for the support and encouragement they have given me right through the completion of this project and my degree.

T   +44 (0)7956 499083
E   amit_ganatra@hotmail.co.uk
W   mysite.orange.co.uk/
     amitganatra

EMMANUEL WICHE

MOBILE COMPUTING BSc

Over the past few years, my technical and creative skills have advanced into new areas such as Mobile Information Device Programming (MIDP), software development, wireless data networks, XML technologies and E-Systems technologies that are based on Java technologies such servlets and JSP's. This advancement in skills motivated me into developing the project shown.

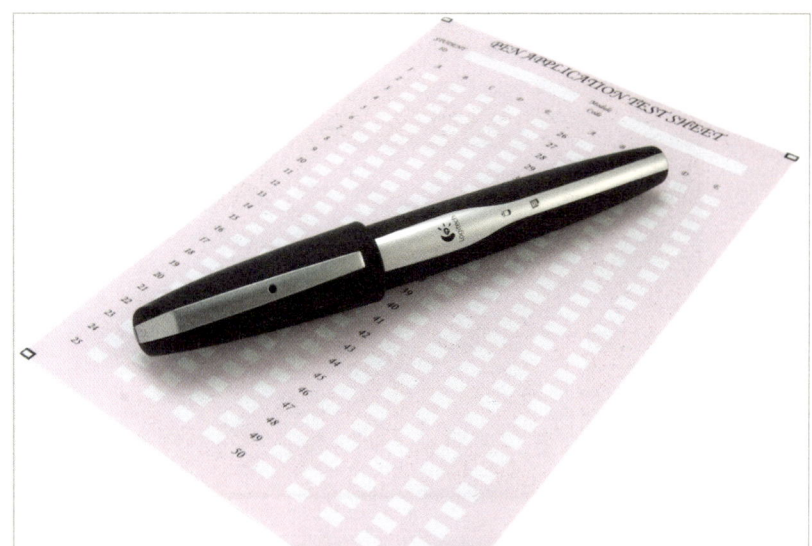

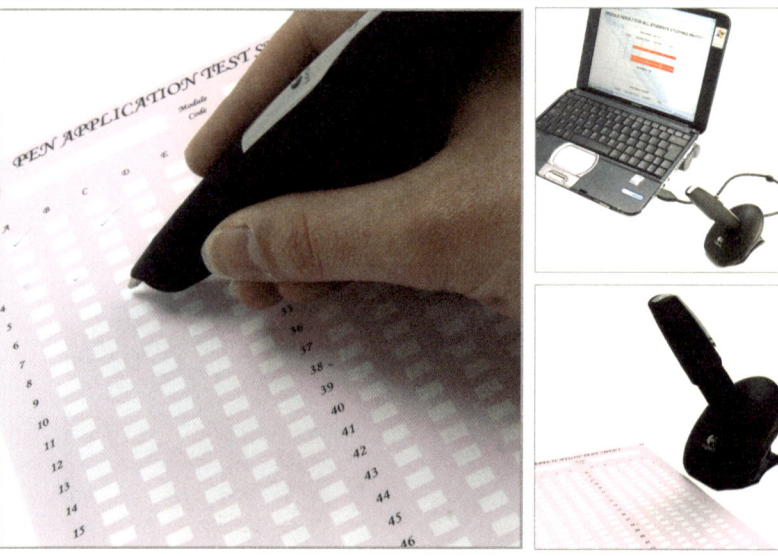

An application which integrates the new portable digital pen and paper technology with Server Side Technologies dubbed "The Digital Pen Application". I hope to use this experience in the future perhaps pursuing a career in web development, software development or wireless networks.

T   +44(0)7984136071
E   ewiche@googlemail.com

TRANSPORT FUTURES

We are a mobile society. Our reliance on
transport networks has never been greater.
The systems to keep people moving,
whatever their destination, need to be highly
safe, reliable and flexible. Overcoming the
technical challenges ahead is in the hands
of the next generation of problem solvers.

SAMUEL BAIRSTOW

INDUSTRIAL DESIGN
AND TECHNOLOGY BA

Transport Futures

A placement year at DEMAND – Design and Manufacture for Disability, has changed my perspective on superfluous design. The work displayed here exemplifies my progression in designing innovative products for the bigger picture. This in turn rewards me with a greater sense of achievement than designing the average consumer product. This course has inspired my outlook on design. I am a fun loving and adventurous individual who thrives off any challenge, design-based or otherwise.

T   +44 (0)7738 220307
E   scrbairstow@yahoo.co.uk
W   www.sambairstow.com

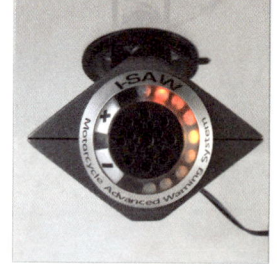

I-SAW Intelligent Situation Awareness; because drivers generally don't see motorcyclists. Just 1% of vehicles on the road are motorcycles and yet motorcyclists make up for 20% of road fatalities. A large majority of those fatalities are attributable to other vehicle drivers lacking awareness. I-SAW uses Radio Frequency transceivers to increase the public awareness of motorcycles on public roads.

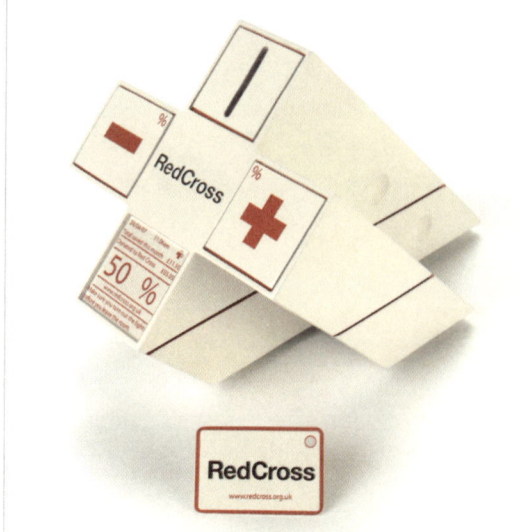

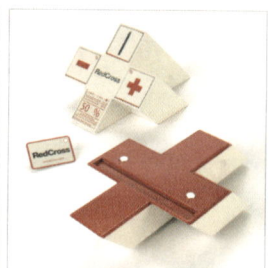

Red Cross Moneybox a future product that stores loose change and donates half of the monthly collection to the Red Cross via a wireless Internet connection. The other half can be stored on your card and spent on what you desire. The Crickaider made for Josh, who has Cerebral Palsy, to improve his cricket and posture. Since using the Crickaider, he has received an award for the player who had shown most improvement.

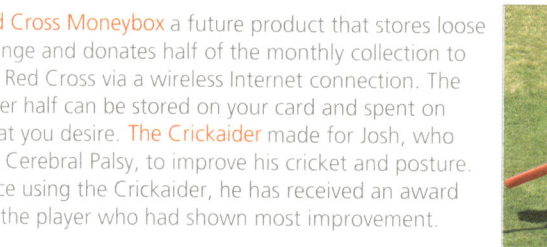

— Transport Futures ——

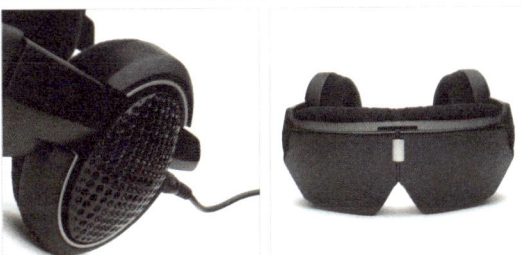

Interactive Adobe Flash Portfolio Website and associated personal branding.

Jet Lag Headset, reducing circadian discord (Jetlag) through a bilateral approach. Firstly, in-flight light therapy realigns the users circadian rhythms, and secondly, ANC and acoustic design reduces cabin noise; resulting in greater productivity and wellbeing upon arrival. Volumetric prototype and fully functioning dosing circuitry are shown here.

The expansive skill set developed whilst at Brunel has been built upon with a years experience in a busy design consultancy. As a fully contributing member of a small design team and has, through freelance experience, culminated in a broad synthesis of skills applicable to any project, also allowing my personal interest in photography to become a professional asset. The application of complementing technology, innovative engineering, and a focus on the interrelationship between the user and product is what drives me.

T +44 (0)7793 019826
E david@davidaconnell.co.uk
W www.davidaconnell.co.uk

MICHAEL
CUNNINGHAM

INDUSTRIAL DESIGN BSc

During my four years at Brunel I took the opportunity to study at San Francisco State University. I took marketing and rapid visualization to widen my skill range. It also helped me to understand the importance of working with different cultures. I hope that in the future I can utilise the skills I have acquired in a range of challenging projects.

T   +44(0)7841 618555
E   michael.cunningham07
     @googlemail.com

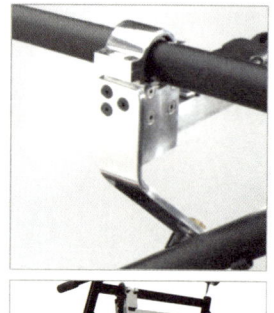

Bicycle Cantilever System is designed to aid the transportation of large objects on a bicycle frame to increase the use of bikes in everyday life for environmental reasons. The objects lie across the back of the bike and on top of the cantilever. The product is compatible for a range frame size and fits onto a standard style frame.

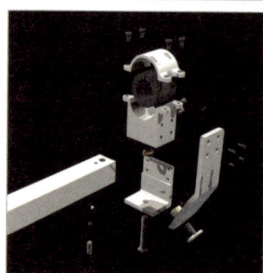

Yakult Desktop Electronic Photo Frame transmits pictures of your family members from your computer and displays a slide show in the head screens. The bodies of the people come apart and can be reversed to change the gender of the people to match that of the family member. There are 3 different sizes for each age group (adult, adolescent and infant). This product is designed to emotionally reassure the user, in keeping with the Yakult brand.

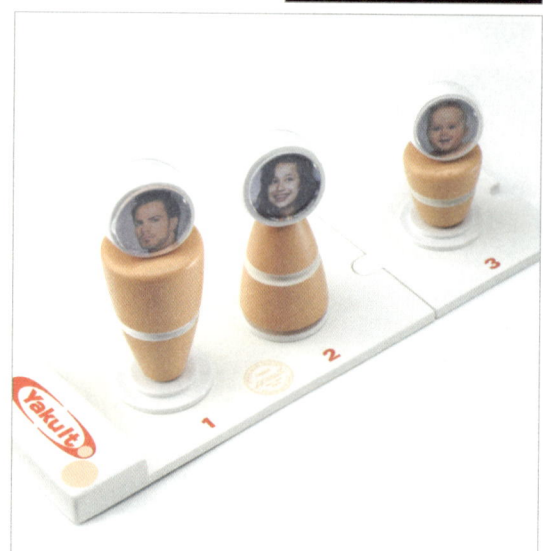

1

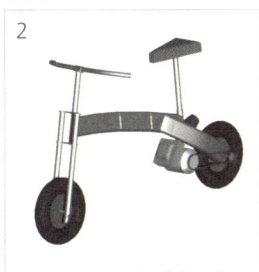

2

1, 2 Foldable motor scooter that combines the advantages of existing methods of commuting. It enables the user to utilise the speed of the suburban train connections while still maintaining the mobility of being motorized inside city boundaries. It is foldable into a trolley during train journeys and has the characteristics of a 50cc motor scooter while deployed.

3

4

3 This is a contextual design project for supermarkets exploring world foods. Recipes for traditional meals are printed on paper bags containing the needed ingredients. Authentic cooking accessories such as tagines would also available for each recipe.
4 A twisted acrylic sculpture.

CASPAR KNEIP

MECHANICAL ENGINEERING
AND DESIGN MEng

◾ Transport Futures

I began at Brunel studying a pure mechanical engineering degree. Because I wanted to be creative and innovative as well as logical and analytical, I switched to the Mechanical Engineering and Design (MEAD) Masters course at Brunel. I am now finishing the five-year thick-sandwich course, which included a work placement year in Germany. Next I will be studying a Masters in Advanced Mechanical Engineering at Imperial College, London.

T   +44 (0)7930 535954
E   casparskneip@hotmail.com

NEIL MEDLER

MOTORSPORT
ENGINEERING BEng

I complemented my studies with a year in industry, where I worked as an Aerodynamic Model Designer with WilliamsF1 Grand Prix Engineering. Time here was spent on CAD model design, wind tunnel testing and engineering drafting. The year in work allowed me to put into practice the theory learnt in the first two years of study, enabled me to build a network of contacts in the motorsport industry and better prepare me for my final year of studies.

My project as part of the Brunel Racing team was to design the rear bodywork and sidepods for this year's Formula Student entry, BR-8. A brief was drafted to highlight areas of importance during the design stages. A low drag coefficient and effective cooling of the radiator cores was critical to producing an effective design. This was achieved by influencing the speed and direction of internal airflow, and was verified with CFD analysis using FLUENT.

Quick and easy access to the internal systems was also highlighted as a key area, along with aesthetic appeal. The final design meets all the criteria set out beforehand, along with a great weight saving on the body panels compared to previous designs.

T   +44 (0)7748 013871
E   neilmedler@hotmail.com

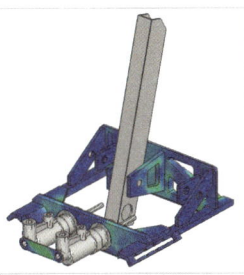

A pedal box provides the control link between the driver's feet and the throttle/clutch/brakes. The BR-8 pedal box marks Brunel Racing's first use of composite-polymer pedals, utilising carbon fibre and Delrin along with advanced bonding techniques. Quick releases offer an innovative adjustment system. Designed entirely using SolidWorks CAD, pedal and frame parts optimised using FEA stress analysis with Abaqus CAE.

KHIMANI MOHIKI

MOTORSPORT
ENGINEERING BEng

- Transport Futures

As an aspiring mechanical design engineer, this project provided the ideal platform to put engineering theory into practice, I was solely responsible for the design and manufacture of the pedal to be used on the Brunel Formula Student car.

I am hoping to continue my studies into the automotive applications of Finite Element Analysis /Computational Fluid Dynamics at postgraduate level and look forward to a career within the automotive industry.

T   +44 (0)7861 396251
E   khimani_mohiki@hotmail.com

LLOYD SCHROEDER

INDUSTRIAL DESIGN BSc

— Transport Futures —

Spending 12 months at Tsunami Axis – the largest supplier of Herman Miller furniture in the UK – has provided me with a solid grounding in the area of project management for design.
At each stage in the design process, my skills were utilised and enhanced, allowing me to enter my final year with the knowledge and ability to produce quality work from paper sketches through to the final model.
Skills: Autodesk 3ds Max, Adobe Photoshop and Illustrator, AutoCAD/MillerCAD, Pro/ENGINEER and modelmaking.

T   +44 (0)7961 989301
E   lloydschroeder@hotmail.co.uk

1 – 4 A bicycle hire scheme for Transport For London that allows the user to activate a bicycle lock with their Oyster card, removing and returning the bicycle with ease at any docking station located around London.
5 This product is part of Prada's new Fashion Technology brand, and uses Radio Frequency Identification technology to make payments and gain exclusive access to events.

BRUNEL RACING

MOTORSPORT
ENGINEERING MEng

Transport Futures

Brunel Racing is a university-based motorsport team, comprising motorsport engineering and mechanical engineering students.

Every academic year the team is tasked with designing, managing, manufacturing and racing a single seat Formula Student racecar.

BRUNEL RACING

MOTORSPORT
ENGINEERING MEng

At Brunel we compete in two major Formula Student events during the summer months. The first of which is the UK Formula Student event at Silverstone, organised by the Institute of Mechanical Engineers. The second event we compete in is the German Formula Student competition. We also take part in various other motoring events such as the Silverstone Learn to Win event, the Autosport show and the Goodwood Festival of Speed.

This years car, BR-8, is an evolution of concepts from previous year's successes. Comprising a steel tubular space-frame, with a rear transversely mounted Yamaha R6 supercharged engine, running on the bio-ethanol blend E85. The car is also equipped with a front and rear mono-shock suspension system, a push button gear change mechanism and a full aerodynamic package.

BR-8 will be competing in two Formula Student competitions during summer 2007 at Silverstone, UK between 12 – 15 July and at Hockenheim, Germany between 8 – 12 August.

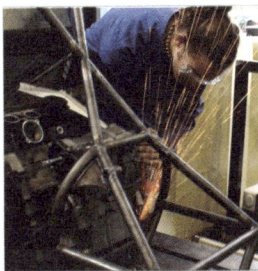

BRUNEL RACING STUDENT MANAGEMENT GROUP

Sam Davies
Kwaku Gueye
Andrew Lee
Dan Marshall
Khushaal Sharma
Gaspar Vieira Leite

BRUNEL RACING TEAM MEMBERS

Ashley Bradford
Phillip Broad
Paul Dance
James Downes
Richard Drury
Richard Hull
Jon Isherwood
James Julian
Tracey Kirk
Costas Kyriacou
David Lane
Christoforos Loizides
Jane Mackay
Neil Medler
Khimani Mohiki
Matthew Payne
Matthew Platings
James Stuart
Simon Swizinski
Jasdeep Riarh
Anish Varsani

BR-8 TECHNICAL SPECIFICATION

Engine
2005 Yamaha R6
4-cylinder 600cc

Differential
Quaife chain-drive LSD

Suspension
Front and rear mono-shocks with V-Keel

Chassis
T4130 steel tubular space frame

Fuel
E85 Bio-Ethanol blend

Tyres
Hoosier 6"/18"-10"
SOFT compound

Induction
Rotrex C15-60
Supercharged System

Body
Carbon-fibre aerodynamic skins

Weight
Under 200kg

Brakes
AP Racing 4 pot
CP4227 callipers

Controls
Custom push button
gear selector

Engine Management
Motec M800 ECU

Lubrication
Custom wet-sump

ARCHIMEDES JET

MECHANICAL ENGINEERING WITH AERONAUTICS MEng
AEROSPACE ENGINEERING MSc

Transport Futures

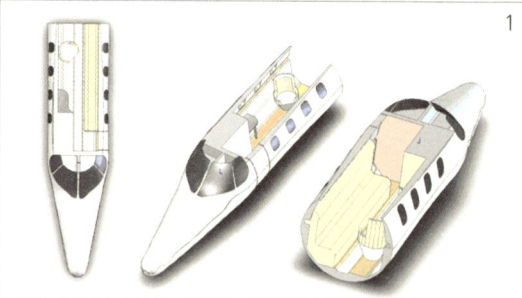

Ewan Browell, Geraint Prisk, Bjorn Madsen, Edward Fisher, Ciaran O'Reilly

This project details the conceptual design of a Very Light Jet (VLJ) aircraft, to compete competitively in the new, emerging VLJ market. It has to comply with CS – Chapter 23 (FAR-23) regulations, and must be a state-of-the-art design. The testing and evaluation of the conceptual design must also be documented.

In order to achieve the objectives set out in the brief, a number of different engineering tools have been utilised. These include Darcorps Advance Aircraft Analysis, XFLR5 3D modeller, SolidWorks and the Brunel University Merlin Flight Simulator.

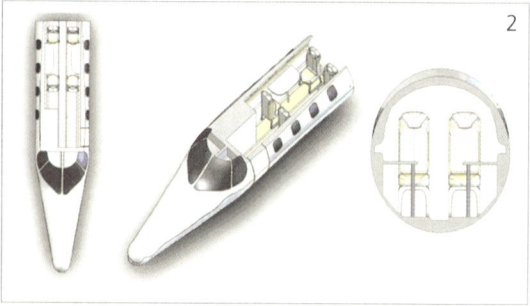

The main design was produced using the AAA program, which is the industry standard aircraft design, stability, and control analysis software. The program was used to rapidly develop the aircraft's configuration from early weight sizing through to detailed performance calculations, whilst always working within the regulatory constraints.

The final conceptual design was tested using the Merlin Flight Simulator. The design was confirmed as the 'best student-designed simulated aircraft' by the test pilot (Dr Guy Gratton). Further recommendations to enhance the aircraft performance were made.

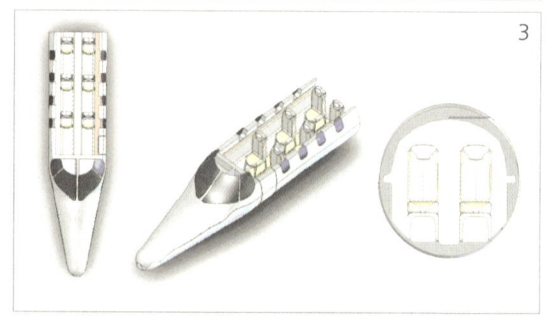

Our design is to be entered into the international Merlin Simulator competition where it will compete against alternative aircraft designs produce by other universities from around the world.

Ewan Browell
T   +44 (0)7967 606501
E   ehbrowell@hotmail.com

Geraint Prisk
T   +44 (0)7809 505949
E   lordoffinity@hotmail.com

Bjorn Madsen
T   +44 (0)7903 104053
E   bjornamdi@gmail.com

Edward Fisher
T   +44 (0)7739 121749
E   er_fisher@hotmail.com

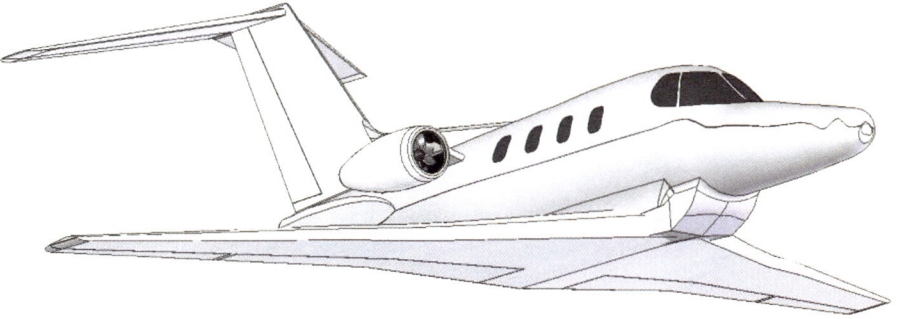

ARCHIMEDES JET

Conceived in 2007, the Brunel Archimedes Jet is a conceptual design of a multi-role Very Light Jet aircraft. It is designed to compete in the business, commercial and leisure markets, offering high performance and range with a relatively low initial outlay and maintenance cost. Primarily designed to carry five passengers the Archimedes Jet can be fitted out to carry six at the expense of luggage space. The Archimedes Jet also has the potential for single pilot operations, thus further reducing potential operating cost, although a conventional two pilot layout is provided. Initial simulations have suggested excellent handling characteristics and a high level of responsiveness.

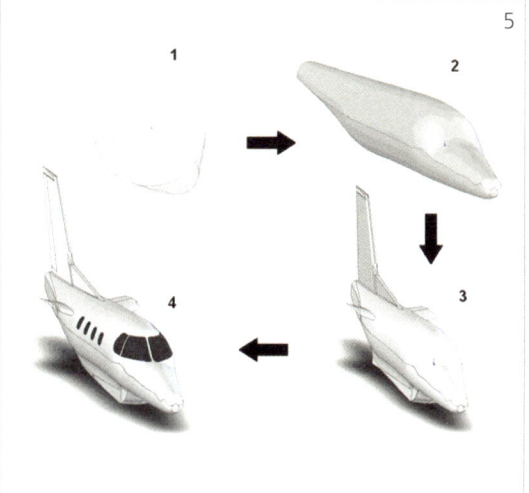

4

1 Private Jet layout
2 Business Jet layout
3 Air Taxi layout
4 Vortex generation
   at wing tips
5 CAD design process

5

Ciaran O'Reilly
T   +44 (0)7904 701911
E   cj_oreilly@hotmail.com

PERSONAL SAFETY
AND SECURITY

In a world fraught with both seen and
unseen dangers, innovators that seek
to protect us without encumbering our
lives, and offer us discreet yet effective
solutions will achieve widespread success.

JAMES ADIDE

INDUSTRIAL DESIGN
ENGINEERING BSc

Personal Safety and Security

My work placement year spent at the Home Office Scientific Development Branch gave me a better insight into the design approach to solving engineering problems. The wealth of knowledge gained from Brunel Design and the work placement have helped prepare me for challenges I will face in the engineering and design fields. My skills include SolidWorks, Autodesk Inventor, Pro/ENGINEER, AutoCAD, ANSYS, MSC Adams, Adobe Photoshop, Illustrator and physical modelmaking.

T   +44 (0)7969 472116
E   james.adide@gmail.com

Sponsored by the Home Office Scientific Development branch, this project involved the design of a multi-axis table that would be used for body armour stab testing in conjunction with the Body Armour Standard for UK Police.
1 – 3 Auxiliary views, giving a representation of the general dimensions of the table
4 The final model, positions of parts and selected mechanisms.

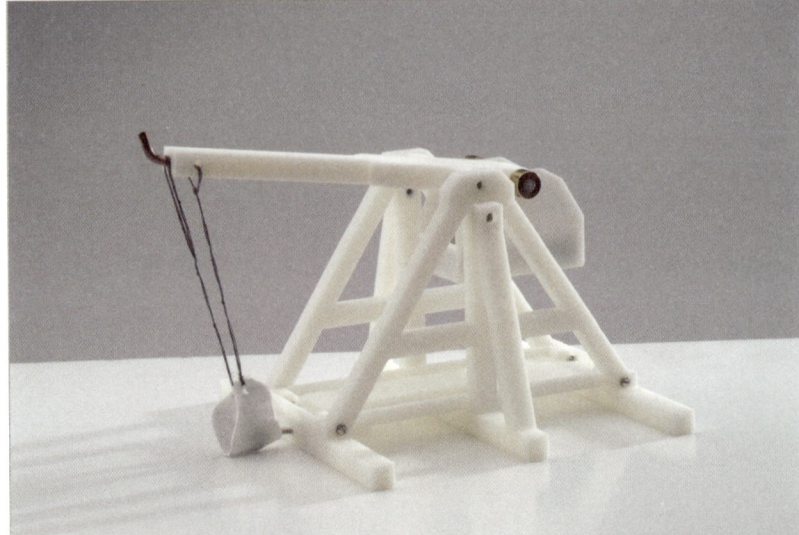

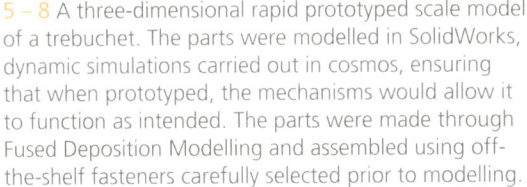

5 – 8 A three-dimensional rapid prototyped scale model of a trebuchet. The parts were modelled in SolidWorks, dynamic simulations carried out in cosmos, ensuring that when prototyped, the mechanisms would allow it to function as intended. The parts were made through Fused Deposition Modelling and assembled using off-the-shelf fasteners carefully selected prior to modelling.

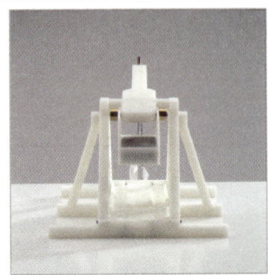

ADRIAN DEFENDI

VIRTUAL PRODUCT DESIGN BSc

- Personal Safety and Security -

Man Overboard Inflatable Launcher designed to fire its inflatable projectile within a range of 60 meters. Once launched, it inflates to an area 3x3 metres that aids in the recovery of multiple people overboard. Designed and developed for use by the RNLI. The M.O.B. inflatable launcher, that incorporates a laser range finder and winch that retrieves the inflatable and persons overboard.

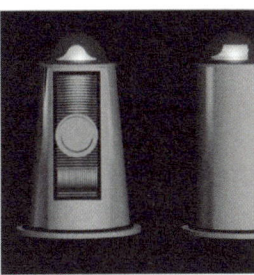

Having studied design for four years I have gained much knowledge and experience from this course. My strengths include a keen knowledge of design software packages including advanced AutoCad, Adobe Illustrator, Photoshop, Pro/ENGINEER Wildfire, 3D Studio Max and scanning software such as Geomagic Studio and the motion capture software post-processing tool EVaRT.

3D Studio Max rendering of architecture study. Rendering from Geomagic Studio 8 of a scanned statue that has had substantial post-processing to create the final cleaned up 3D model.

T   +44 (0)7946 416560
E   defendi.design
    @googlemail.com

PHILIP DOWNTON

**INDUSTRIAL DESIGN
AND TECHNOLOGY BA**

Personal Safety and Security

During my three years at Brunel University, I have developed an understanding of how design can play such a critical role in our lives. I am more aware of the impact product design has upon the environment and I am very passionate in working towards a sustainable future, a passion which is evident throughout my work.

Skills: modelmaking, Adobe Illustrator, Photoshop and InDesign, SolidWorks, Pro/ENGINEER, AutoCAD, Autodesk AliasStudio, Adobe Dreamweaver, HTML.

T   +44 (0)1872 510026
E   philip_downton@
    hotmail.co.uk
W www.philipdowntown.co.uk

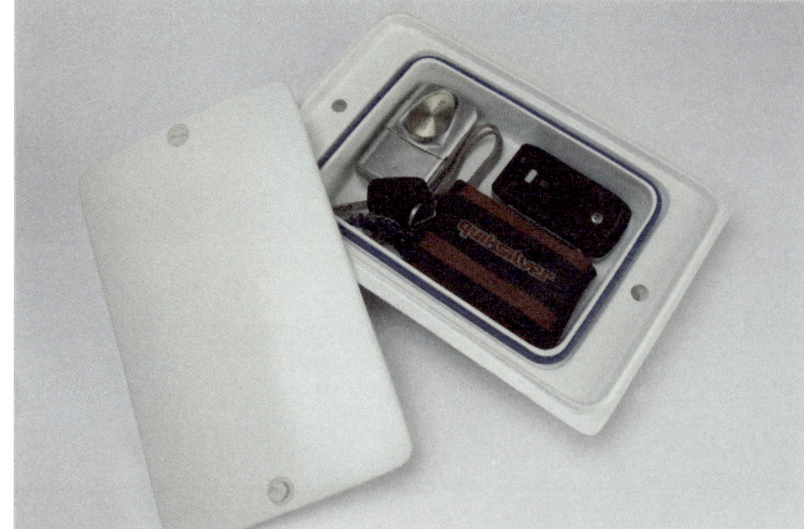

Surf Safe is a concept designed to give the user the opportunity to store their most valued and essential possessions whilst surfing. Key features include a watertight seal, quick release opening and closing system and an excellent strength to weight ratio. Surf the waves in the knowledge that your possessions are safe in your board.

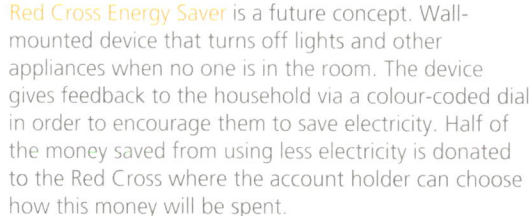

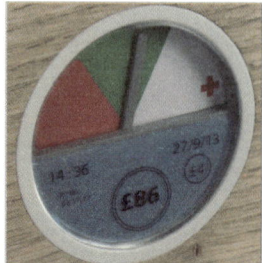

Red Cross Energy Saver is a future concept. Wall-mounted device that turns off lights and other appliances when no one is in the room. The device gives feedback to the household via a colour-coded dial in order to encourage them to save electricity. Half of the money saved from using less electricity is donated to the Red Cross where the account holder can choose how this money will be spent.

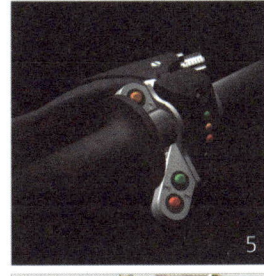

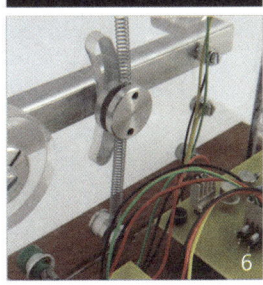

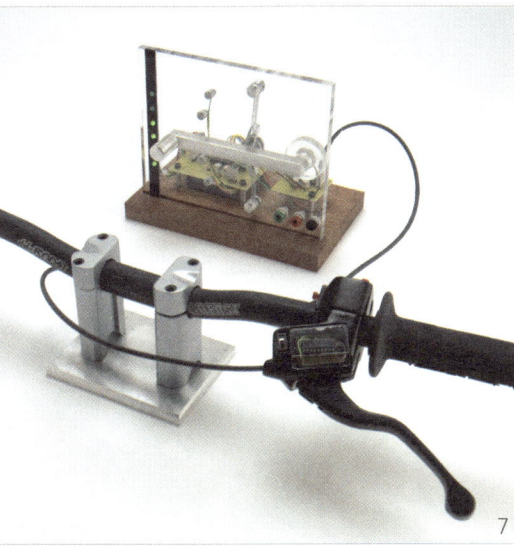

SIMON ELLIOTT

PRODUCT DESIGN BSc

- Personal Safety and Security

Studying Product Design at Brunel has allowed me to develop my skills to a level where I am ready to begin designing in industry. I have learnt that design can have broad boundaries and that often the simplest of sketches can produce the foundations for the design of a successful product. Having spent a year in placement at Lumitron Lighting I have also developed my ability to communicate ideas to both clients and colleagues, as well as experiencing design in a real environment.

1 An onboard lap timer for use in amateur motorsports. 2 Screenshot of my online portfolio developed using Macromedia Flash. 3 Sample calendar page produced to add function to my photography. 4 A simple brochure designed as a handout for use during a presentation. 5 – 7 Handlebar-mounted motorcycle gear lever with gearbox prototype used to show the functionality and CAD rendering of the final product created in SolidWorks.

T   +44 (0)7796 265184
E   simelliott@hotmail.com

FENELLA HOLDEN

INDUSTRIAL DESIGN BSc

I believe in design that results in simple, contemporary solutions to complex problems; where details exist for a purpose and style transcending the demands of fashion.

Placement: CHANEL Design London, 2005-2006

Skills: Adobe Photoshop, Illustrator, InDesign, AutoCAD, as well as HTML and JavaScript coding, rapid hand rendering, sketching, model making and photography.

T   +44 (0)7841 374449
E   fenella.holden@ googlemail.com

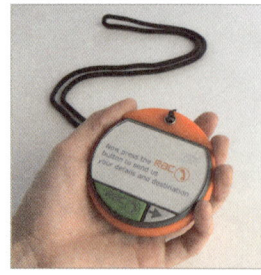

The SOS Guardian is a concept for the RAC that makes the breakdown experience much safer and less traumatic. Compactly stored behind your tax disc, this reassuring product continually updates how long you will have to wait for help to arrive when broken down. Using GPS technology it sends the exact location of your vehicle to the breakdown company, it then informs you with easy step-by-step instructions where to wait in the best position of safety.

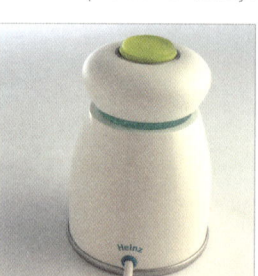

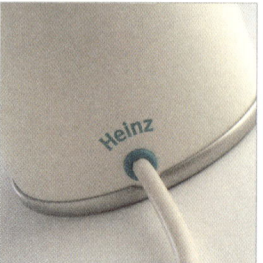

Heinz 'Journeyz Friend' is a future concept, using Heinz brand family values. The product allows subtle communication between parent and child whenever the child goes on a journey. The clip on the child's bag sends a message to the parent product when the child arrives at school. By activating the parent product a reply is sent to the child.

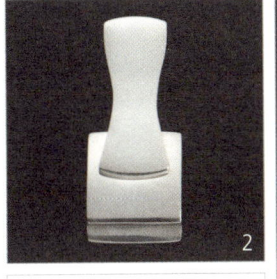

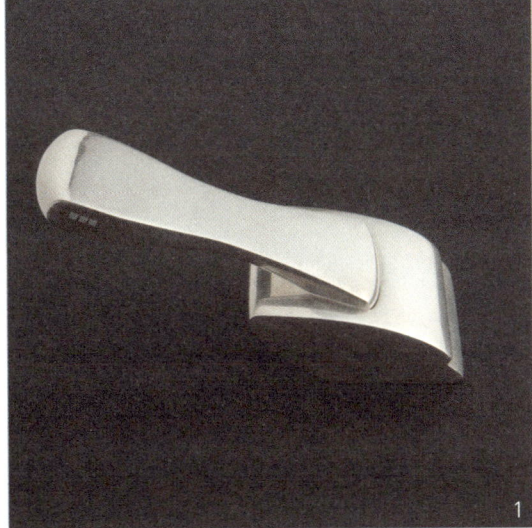

DANIEL NOONAN

INDUSTRIAL DESIGN BSc

— Personal Safety and Security —

One of the core principles of industrial design is to understand a subject as deeply as possible. To that end, it is my aim to have an excellent knowledge base in all my areas of interest. I think that history and philosophy are important because it is only by knowing what doesn't work and the history behind a subject that we can move forward; context is king.

1 – 3 Prada Key - Solid silver wireless key. The brief for this project was to develop a product for Prada that combined technology and fashion. 4, 5 Strata - Second year lighting project, an outdoor light that is turned on by passing a hand through a beam of light. 6, 7 Hand held temporary tattoo printer - Major Project Cartridge design for use in hand held printer, combining ink, battery and data into one form.

T   +44 (0)7905 518687
E   noonan_dr@yahoo.co.uk
W   www.dnoon.com

BEN SILLENCE

PRODUCT DESIGN BSc

— Personal Safety and Security —

With an experience within the industry of taking designs from concept to shopfloor, and a desire to produce creative and elegant solutions, I have been inspired and filled with enthusiasm by my time at Brunel – not only by professionals, but most importantly by my peers. This has encouraged me not only to design products that are creative in their solutions, but also that embody individualism and personality.

T   +44 (0)7810 821186
E   ben@thekickinside.co.uk
W   www.thekickinside.co.uk

Sydney The Line Following Truck uses a micro-controller system to bring the large scale of the construction site to the home by following a black track laid down by the user. Sydney is made entirely of wood reclaimed from the old Runnymede campus, reducing the environmental impact of manufacturing the product.

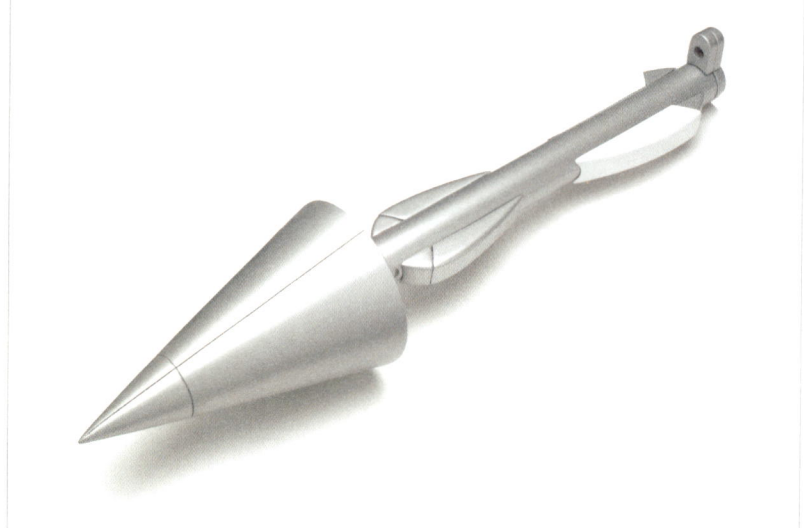

SandSafe is a simple and elegant solution for protecting ones possessions against theft while on the beach. Once pushed into the sand, SandSafe increases its surface area and will resist a pulling force of over 30kg, providing an 'anchor' to which a steel net, containing the user's possessions, is padlocked. A tamper-proof mechanism allows the user to retrieve the product from the sand and reuse it.

PAUL TINKER

PRODUCT DESIGN BSc

I have always had a great passion for drawing and rendering, and I seek to integrate them into any project. I feel that they are the most successful methods of communication in any context. My placement at Gillette gave me an excellent opportunity to both collate and apply the skills learnt at Brunel to a wide variety of projects; one of which included design from concept to prototype. The experience has vastly furthered my understanding of all aspects of the design process, as well as protocol in a professional environment.

1 Foldable cycle helmet design. 2 Foldable helmet prototype that was tested under the BS EN 1078 standard by BSI. 3 The helmet folding mechanism. 4 The helmet segments are locked together by the action of the user pulling on the straps. 5, 6 A pencil mini-gun that uses voice recognition to fire the desired coloured pencil. It can be reprogrammed by the user. 7 An image taken from a themed, self-branded website.

T   +44 (0)7834183762
E   paultinker@hotmail.com
W   www.idea-depot.net

LAURA WILLIAMS

INDUSTRIAL DESIGN BSc

Personal Safety and Security

I love to create, be it music, clothes, products or dance. All forms of creativity allow me to be expressive and indulge in a personal passion of mine; which is to be more than just superficial. Intertwining products, with a more meaningful user experience, is an exciting way forward for design and society, and it is an area of design that I look forward to contributing to.

T   +44 (0)7941 087664
E   ldiablas@hotmail.com

Safe Knife was designed in response to both the growing issue of knife crime in the UK, and to the safety issues associated with dropping a kitchen knife. The blade is designed to retract into the handle once moved beyond a certain level of acceleration, whilst the inclusive handle is comfortable for both arthritis and non arthritis sufferers to hold.

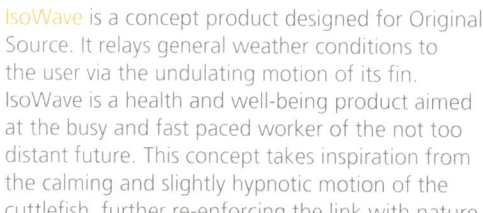

IsoWave is a concept product designed for Original Source. It relays general weather conditions to the user via the undulating motion of its fin. IsoWave is a health and well-being product aimed at the busy and fast paced worker of the not too distant future. This concept takes inspiration from the calming and slightly hypnotic motion of the cuttlefish, further re-enforcing the link with nature.

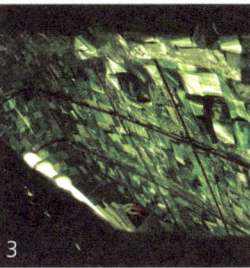

1

2

3

4

1, 2 neenaw is an emergency vehicle detection system for deaf and hard-of-hearing drivers, designed in conjunction with the RNID. It 'listens' for emergency vehicle sirens and warns the driver when they approach. 3, 4 SNAK is a light made entirely from recycled crisp packets, used to raise awareness of environmental issues through lighting. 5 Touch is a futuristic concept for Dove, from which the feel and warmth of a human touch can be sent and received.

HELEN WRIGHT

INDUSTRIAL DESIGN BSc

I loved every second of my time at Brunel and am now ready to move on and see where life takes me. This course and my experience from a year's placement at Paperdog, a graphic design company, and my role as Creative Team Leader for MADE IN BRUNEL have made me passionate about design. When I'm not working at my desk I love being outdoors, playing tennis, squash, rock climbing and socialising with friends.

Skills: Mac and Windows user, Adobe Photoshop, Illustrator, InDesign, QuarkXPress and SolidWorks.

T   +44 (0)7730 578439
E   helenlucy_wright@ yahoo.co.uk
W   www.helen-wright.com

RESPONSES TO
CLIMATE CHANGE

We have an incredible impact
upon our environment. To sustain
our growth and development as
a society, we need to consider the
most efficient uses of our resources.
Advanced thinkers in design and
engineering will provide a means
to a maintainable way of living.

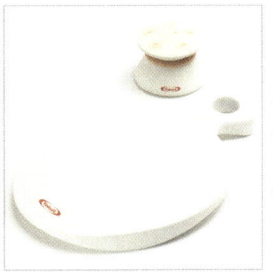

Kokeshi – Conceptual product for probiotic drinks manufacturer Yakult. Kokeshi provides a revolutionary new way of maintaining health through monitoring the food we eat. Taking its name from the Japanese doll that inspires its design, Kokeshi utilises non-invasive ultrasound technology to determine the chemical composition of meals and informs the user of daily intakes of fat, sugar, salt, protein and calories.

1

2
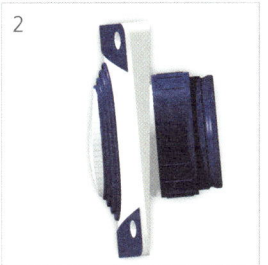

3
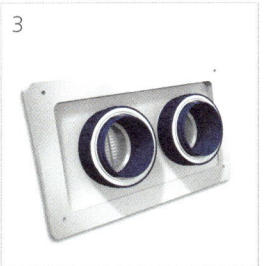

1 – 3 In-taxi Pollution filter. Utilising HEPA filtration technology, the device filters out air-borne pollution, passing air into the cabin of equivalent air standards to those observed in hospital operating theatres.
4 Watch face designed in collaboration with Swiss watch manufacturer ORIS during a 12 month placement with WilliamsF1 Engineering Ltd. The watch reflects the character of one of the F1 pilots and is now in production in a limited run of 2400 units.

4

CARLO BELLI

INDUSTRIAL DESIGN BSc

— **Responses To Climate Change** —

The ever increasing responsibility for designers to inspire change is at the heart of my interpretation of the design process. Only through truly understanding the needs of the user can we truly bring about holistically innovative design. My time at Brunel and my 12 month placement at WilliamsF1 Engineering Ltd have provided me not only with the core skills essential to good design, but have taught me the roles that marketing and branding play and their existence as an intrinsic part of the design process.

T   +44 (0)7955 079474
E   cjb.design@gmail.com

MARIA DOLKA

INDUSTRIAL DESIGN
AND TECHNOLOGY BA

Responses To Climate Change

The three years I have spent at Brunel, combined with a year's placement in the industry, have taught me to approach the design process from different perspectives. With a keen interest in the environmental effect products have and user interfaces, I aim to develop concepts with a problem solving and environmentally responsible approach.

T   +44 (0)7980 053334
E   maria@mariadolka.com
W   www.mariadolka.com

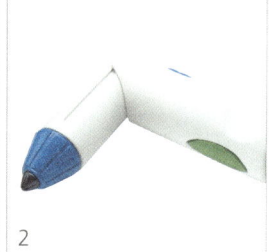

1 A baby intercom monitor detecting pressure in a caterpillar toy. The parent is alerted when the toddler is restless, a strain gauge detects pressure on the toy. Sound and light in the toy is the activated, to help the toddler go back to sleep.

2, 3 Always Power tool. A power tool designed for the increasing requirements for women. Its telescopic design contributes to more effective storage, and its light weight and ergonomically sound form make this product easier to use.

4 – 6 Environmentally Aware Night-Light. This product contributes to reducing the carbon dioxide emissions from households. Powered by solar panels, it is a nightlight that enhances user interaction and increases awareness of the damaging effects to the environment.

5

6

Personal Identity and Logo

I developed a brand and identity that represented me and my work as a designer. The logo's clean-cut image and neatness represents the final outcome of my work while the backdrop represents the passion, dedication and the process that is involved in any piece of work that I undertake.

HENRY ELLIS-PAUL

PRODUCT DESIGN BSc

— **Responses To Climate Change** —

After spending a year in the design industry, I have developed a professional and meticulous approach towards my work. During 8 months working at Crown Technology, an international leader in packaging design, I gained experience in designing for high-profile companies including Nestlé. For a further four months, I worked for the award winning toy company Wow Toys, in which I redesigned one of their products; this redesign was released earlier this year. While on my placement I had work exhibited at the LCE 2006 Conference in Belgium, as part of an international design competition.

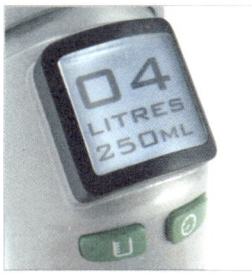

Tap Meter – the aim of the project was to develop a device that would encourage users to minimise their water consumption. The design consists of a domestic tap with an inbuilt display that informs the user of the extent of their usage each time the tap is used. This information changes the user's habits and behaviour through involvement and emotional attachment to the product. This design was exhibited at this year's Ideal Home Show as part of the 2020 Living Competition.

T   +44 (0)7796 498001
E   henryep@hotmail.co.uk

DAVID GADD

INDUSTRIAL DESIGN
AND TECHNOLOGY BA

Responses To Climate Change

My time at Brunel has fuelled my ambition, creativity, and determination and has given me a diverse skill set that I can use to tackle many of the problems in today's rapidly changing and adapting design industry.

Studying abroad at the University of Technology, Sydney and having the role of Marketing Leader for MADE IN BRUNEL has advanced my analytical thinking, time management, creativity and my ability to understand from others' perspectives.

I enjoy solving real problems in design and society and identifying latent user needs not addressed in current products whilst being an environmentally conscious thinker.

T   +44 (0)7947 213737
E   david.e.gadd@gmail.com
W   www.davidgadd.com

A Brush For Life is a modernisation of traditional paintbrushes that aims to add value over conventional brushes, in which sales are usually driven by cost, to gain competitive advantage. The features work to fulfil latent needs of the user and to reduce the negative environmental aspects associated with painting and decorating.

The Yakult Lantern is a conceptual product for Yakult that emotionally reassures the user by displaying information specific to the user's health and vitality. The Lantern then guides and advises the user how to improve their overall wellbeing.

The Colour Amplifier is a product that magically senses any colour it is placed on and shines the same exact colour right back, colours can be captured, stored and converted to a sound.

STEFAN GROSVENOR

INDUSTRIAL DESIGN BSc

Responses To Climate Change

Climate Calm comprises both a base unit and blanket, in which cooling and heating can be achieved to maintain and control a comfortable environment during sleep. Designed as a response to increased night time temperature, as a result of global and industrial warming. Sponsored project, in association with Helium3 Design.

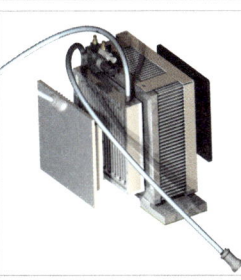

I believe design is the strive for perfection, which is reached when there is nothing left to take away.

My time, both at Brunel and working alongside Helium3 has enabled me to achieve a strong grounding in the design process inherently involved in product development, as well as instilling the importance or appreciation.

I believe that the responsibility for people to change rests not with the people themselves, but with the design of products that can bring about this change.

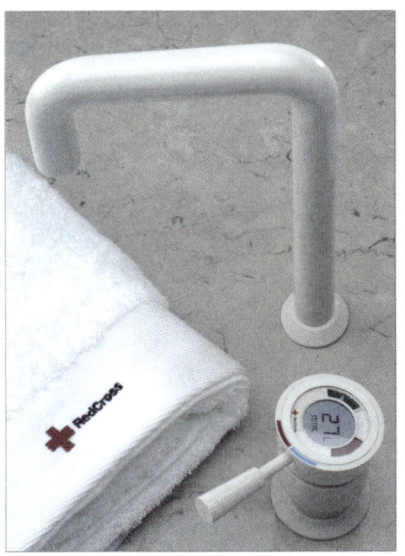

Water and energy saving tap is a future concept for Red Cross, designed to reduce the amount of electricity and water wasted through household water consumption. The product is part of an equation used to both help others less fortunate, as well as educating users with their potential.

T   +44 (0)7813086642
E   sjfgrosvenor@aol.com

JASON LAU

PRODUCT DESIGN BSc

Responses To Climate Change

Experience: 2005 –
San Francisco State University
ERASMUS Exchange
2006 – Web designer at
BetterWorldBooks.com

Skills: Adobe Photoshop,
Illustrator and InDesign,
Pro/ENGINEER, SolidWorks, Adobe
Flash and Dreamweaver, HTML,
AutoCAD, Autodesk AliasStudio,
PIC programming, modelmaking.

Interests: motorcycles,
snowboarding, table tennis
(Club Organiser and Brunel
Team Representative),
fencing (Brunel Team).

T   +44 (0)7786 430432
E   jasonlau128@hotmail.com
W   www.jasonlaudesign.com

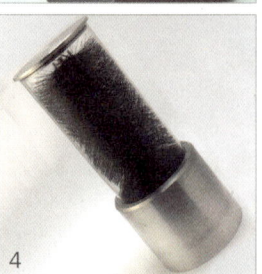

1 Mechatronics Motorcycle Lock with integrated alarm
and infra-red remote control. Operated using tilt switches,
Microchip 16F628, PIC programming and a 24V solenoid.
2 Close up of infra-red remote control. **3** Fold-out Business card. **4 – 5**
Chopstick Washer encouraging the use of reusable chopsticks in Asia.

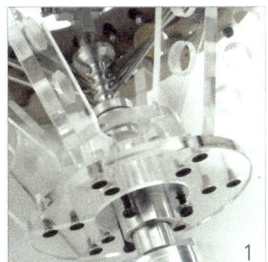

1

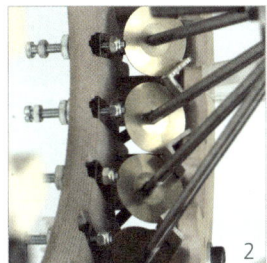

2

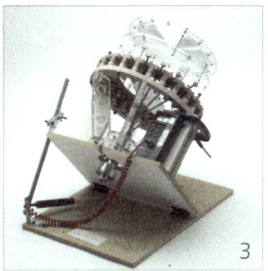

3

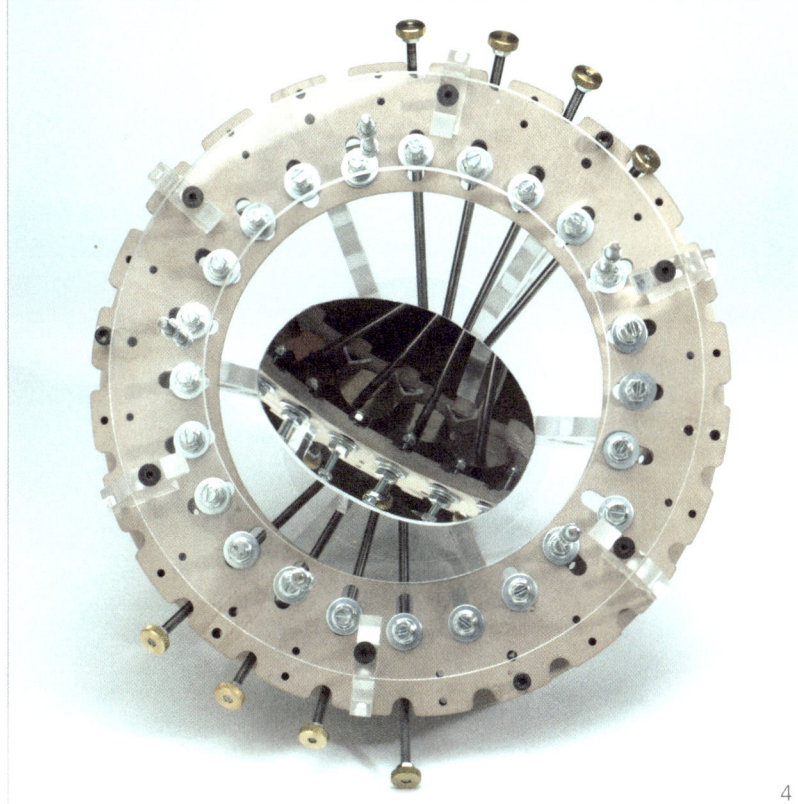

4

THOMAS LEECH

INDUSTRIAL DESIGN
AND TECHNOLOGY BA

— **Responses To Climate Change** —

Having completed a highly stimulating Art foundation course, Brunel has allowed me to build on this creativity with a broad range of professional design skills. A 12 month placement at Morphy Richards has emphasised the value and importance of clear communication when articulating ideas across corporate levels of business.

It has been a pleasure to spend four years working around passionate people who enjoy the challenges of the course whilst being able to appreciate the responsibility that we have in designing sustainable, worthwhile products for our global society.

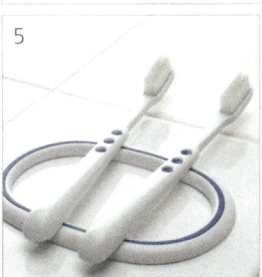

5

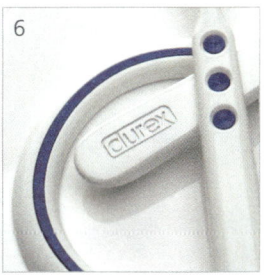

6

1 – 4 This major project is a research study which looks at novel methods of extracting useful mechanical energy from sunlight by utilising the deflection forces of thermostatic bimetals. This iterative process has spawned many development models and computer simulations which have been recorded and assessed in order to assist future research.

5, 6 A future product direction for Durex aiming to integrate the core values of the brand into simple, elegant items which help to support the development of the relationship. Through this product, Durex is identifying the point in a relationship at which a toothbrush may be left at a partner's house as being a symbol of long-term commitment.

T   +44 (0)7988 647864
E   tjleech@hotmail.com

ROBERT MACBETH

MULTIMEDIA TECHNOLOGY
AND DESIGN BSc

- Responses to Climate Change -

As an experienced large-format commercial studio photographer, my knowledge of still images and manipulation have helped me adapt to the interactive multimedia revolution.

Two disciplines I have enjoyed exploring are 2D Animation and video/filmmaking, from editing to publishing. Flash animation and interactive web design (with most pages hand-coded) are what I will take with me to the future.

While on this course I have also learned Object Orientated Programming, allowing my creativity and designs to evolve from a static image to a fluid one. My determination and passion for what I want to achieve might be classed as obsessive by some people, but then who doesn't have an obsession about something?

T   +44 (0)7935 955074
E   robert_macbeth@msn.com
W   www.h2ofacts.co.uk
    www.mac-designs.co.uk

For years I have been watching the amount of water we use/waste in the UK compared to the rest of the world and I discovered the worst water wasters are actually our own country's water companies who are in charge of supplying our water.

I created a Flash animation aimed at children to educate them about the real dangers of wasting water in the UK with an emphasis on our water companies to set a good example.

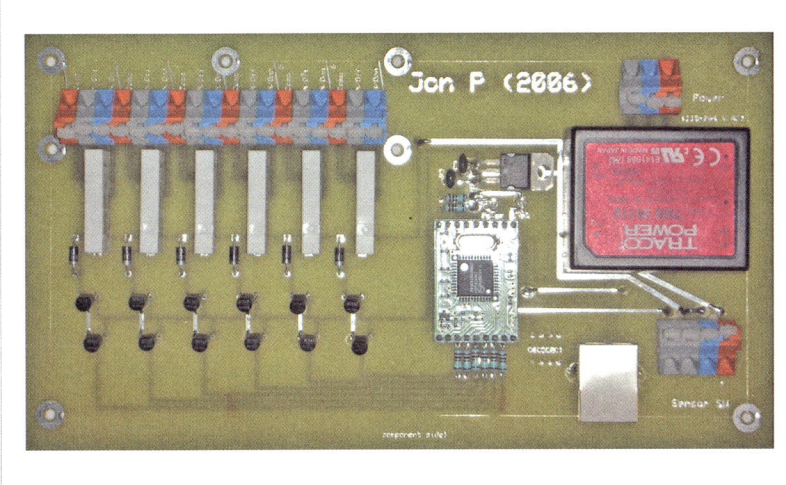

JON PARADI

MECHANICAL ENGINEERING
WITH BUILDING SERVICES MEng

─ Responses to Climate Change ─

Presently the vast majority of commercial and industrial buildings are designed and constructed with a series of intelligent management systems controlling the air conditioning, lighting, security, fire protection, vertical transportation, and electrical power supply.

To make systems within new buildings more cost competitive, manufacturers have taken advantage of the continually dropping prices and the increasing capability of microprocessors to integrate with existing technologies to improve efficiency and functionality.

Older buildings and structures (constructed prior to 1980) seem to only be modernized by necessity rather then to promote energy savings. The market tendency is only to incorporate these intelligent control systems into large developments where the high costs of the control systems can be easily justified.

The past few years at Brunel have been the most rewarding part of my education. I've made some great friends and learned a great deal about subjects that interest me.

Achieving the IMechE project prize was the highlight of the last year for work in level 3.

My main interests include, Archery, Model Engineering and Renewable energy.

I will definitely miss Brunel and remember the time I spent here forever.

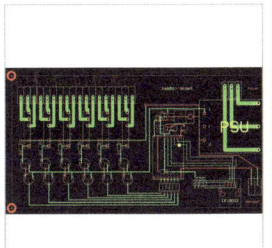

At present there is a market gap between the high cost embedded microprocessor controlled systems used in large modern structures such as skyscrapers, office towers, airports and the often non controlled smaller premises used for small business or in a domestic capacity. The goal is to develop a product that can bridge the gap between the two segregated market divisions.

T   +44 (0)7986 870360
E   jonathanparadi@gmail.com
W   jonathanparadi.
     googlepages.com/main

BEN PAWSEY

INDUSTRIAL DESIGN BSc

Responses to Climate Change

Good design arises from ambitious conceptualisation. Pessimistic design leads to a stagnant, saturated market place. Therefore, designers need to continually ask questions through their work to keep the momentum of progress moving forward. I also believe that utilising sustainable design is an exciting and universal way of ensuring this now and in the future.

T   +44 (0)7792 838440
E   ben_pawsey@hotmail.com
W   www.pawseydesign.com

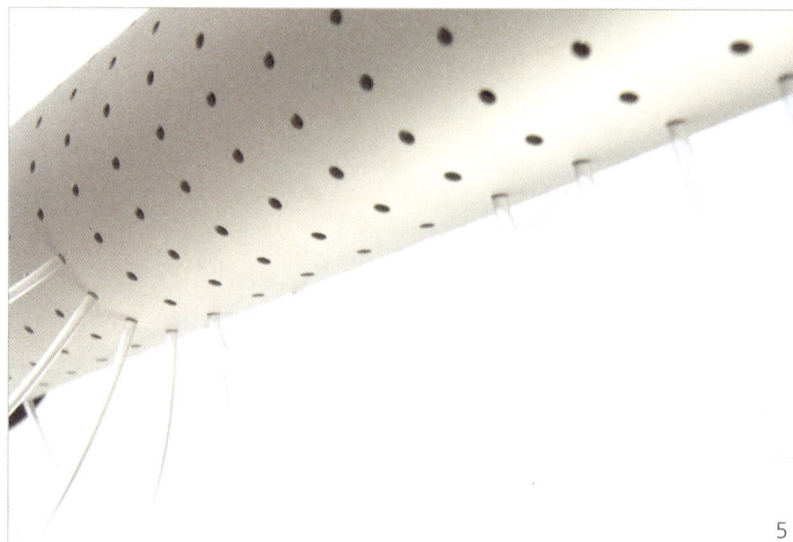

1 Erbie, the herb plant pot that grows in height and diameter with its plant. **2 - 4**. Eden 5 is an earth fertilisation kit that can be broken apart into equipment that will allow people in the developing world to grow a sustainable food supply in infertile soil. **5** i-Rod is a light that allows its owner to personalise its form by plugging fibre optic strands into its array of holes.

1

THOMAS PYNN

INDUSTRIAL DESIGN AND
TECHNOLOGY BA

The course has allowed me to
establish strengths throughout
practical applications in many
areas. I am a driven and
competitive individual with
motivation towards both personal
or group success. I leave Brunel
with the desire to enhance the
practical and methodical skills
learnt and ensure success of
any projects I will work on.

On completing a placement
year at San Fransico State
University I have gained vital
experience in the professional
communication of ideas within
an international group.

2

3

5

4

1 – 2 My leather bound portfolio is a blend of minimal yet
effective design which reflects my current style.
3 – 5 Eco-friendly Mini Bar. A product that can incorporate all the required features
yet cost less to manufacture, use and dispose. The design incorporates Techni Ice
technology to chill the beverages and requires no electricity in this user phase.
Designed for the Travelodge hotel this unit is more flexible and cost effective for
their specific needs.

T   +44 (0)7790 584719
E   pynnster@hotmail.com

TOBY
STEVENSON

INDUSTRIAL DESIGN
AND TECHNOLOGY BA

Responses to Climate Change

My journey through Brunel has developed my skills as a designer and to appreciate the different aspects of the design process, from a simple sketch through to a working product. It has also made me realise what my strengths are, which has encouraged me to pursue a career within the field of rapid prototyping.

T   +44 (0)7886 296875
E   toby_stevenson
      @hotmail.co.uk

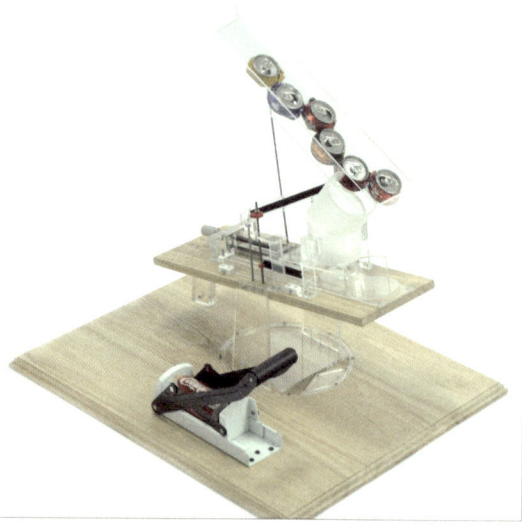

There is a growing consensus that the way we are living is having an affect on climate change and the environment. One way to reduce this problem is to recycle. This project encourages recycling by making it more appealing to people and something that they want to do. The product fires crushed aluminium drink cans as a game to make the collection phase of the recycling process more attractive than throwing the can in the bin.

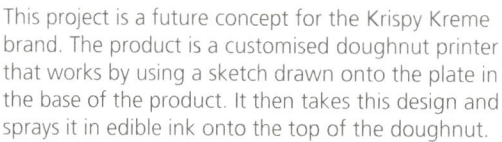

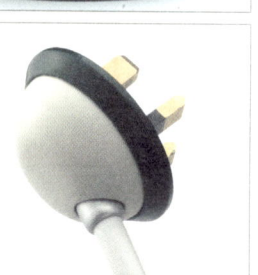

This project is a future concept for the Krispy Kreme brand. The product is a customised doughnut printer that works by using a sketch drawn onto the plate in the base of the product. It then takes this design and sprays it in edible ink onto the top of the doughnut.

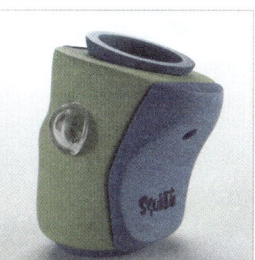

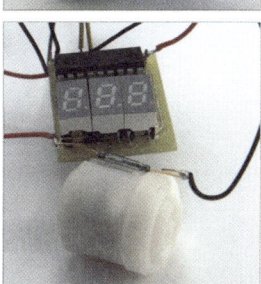

MEGHANA VAIDYANATHAN

INDUSTRIAL DESIGN BSc

— **Responses to Climate Change** —

Squirt At our current consumption rate, it is predicted that we could use up to 40% more water in the next 20 years. Squirt is an awareness-based water meter designed for children aged 3 to 6 and aims to instil conservational etiquette in the mind of a child. Squirt has a child-friendly interface and displays the amount of water consumed over a period of time from the tap to which it is attached.

Four years of experience, both at Brunel and in industry, have been valuable and fundamental in terms of my growth as a designer, and as an individual. I now leave Brunel with a new sense of creativity, technical ability and determination that will provide a solid foundation for any future endeavours I undertake.

Experience: Three Blind Mice, London 2005-2006

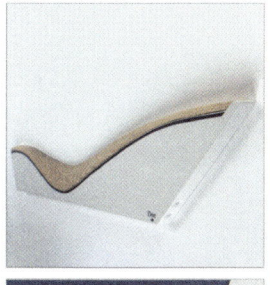

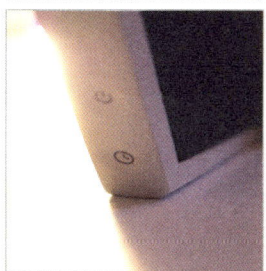

Rise A future concept for Dove, which expands the current market into a product-based one. This wall based alarm clock manipulates light to improve sleep quality and maintain feelings of wellbeing. Rise monitors the user during the night, and wakes him gently, at the most agreeable moment so his day always begins with a refreshing start.

T   +44 (0)7709 579910
E   meg.vaidy@gmail.com

DOMESTIC COMBINED HEAT AND POWER FUEL CELL SYSTEM

MECHANICAL ENGINEERING MEng
MECHANICAL ENGINERING WITH BUILDING SERVICES MEng

Responses to Climate Change

Simon Gilbert

The broad experiences I have had the opportunity to be a part of, including two years on the MADE IN BRUNEL project have changed me in ways I could not have previously envisaged. I have become a competent engineer, with a wide range of skills, including numerical analysis and systems design as well as a solid grounding in key engineering disciplines. 18 months commercial experience have given me confidence in an industrial environment.

Natalie Morris

I am an active, passionate and motivated person towards every aspect of my life. From a young age I have had experience in the field of building services engineering visiting a variety of sites with my father. As I grew, my interest in this field grew with me. Building services accounts for approximately 60% of the world's energy consumption. This being a key area for my passion since there is significant scope for energy savings, which is what I hope to

Building services design is governed by UK and EU legislation as a result of concerns over the rate of energy consumption, fossil fuel depletion and climate change. The use of renewable and sustainable energy technologies in buildings is becoming more viable and its use has increased. This project focuses upon a renewable Combined Heat and Power (CHP) application suitable for a modern domestic dwelling.

We aimed to provide a developed design for a renewable energy CHP system for a single domestic dwelling, ready to be prototyped. The system should made use of hydrogen-based fuel cells, as well as using solar power in the summer to increase overall system efficiency

Fuel cells are electrochemical devices, converting hydrogen into water, producing both electricity and heat simultaneously. Fuel cells use hydrogen (H_2) which is readily available, although the majority of H_2 is steam-reformed using fossil fuels, H_2 can be produced through electrolysis, via more environmentally sustainable methods. Maximum Cell CHP-operation efficiencies of fuel cells are between 60 – 85%, depending on the fuel cell type. During use, fuel cells produce virtually no noise or other harmful pollutants, such as CO_2 or NO_x

The project began with developing a comprehensive virtual model of a typical dwelling for the system, incorporating the latest building regulations and design guidelines. Prospective fuel cells types were evaluated, and the most appropriate selected. Using degree-day data, it was possible to then size the system both in terms of fuel cell capacity and area of roof-mounted solar photovoltaic cells, for optimum efficiency and energy supply to the dwelling.

The system was evaluated with regards to user health & safety, using Risk Management techniques. Environmental suitability was assessed using a bespoke toolset based upon the widely-used Life Cycle Assessment, and economic viability was also determined using a financial model which incorporated future energy price trends.

Recommendations from the project suggest that fuel cell-based CHP with solar energy extraction is an attractive viable environmental alternative. Economically, however the system is more suitable to larger residential applications, such as university halls, or hospitals. This project has allowed the team to hone their skills as mechanical engineers, and was a truly fulfilling learning experience.

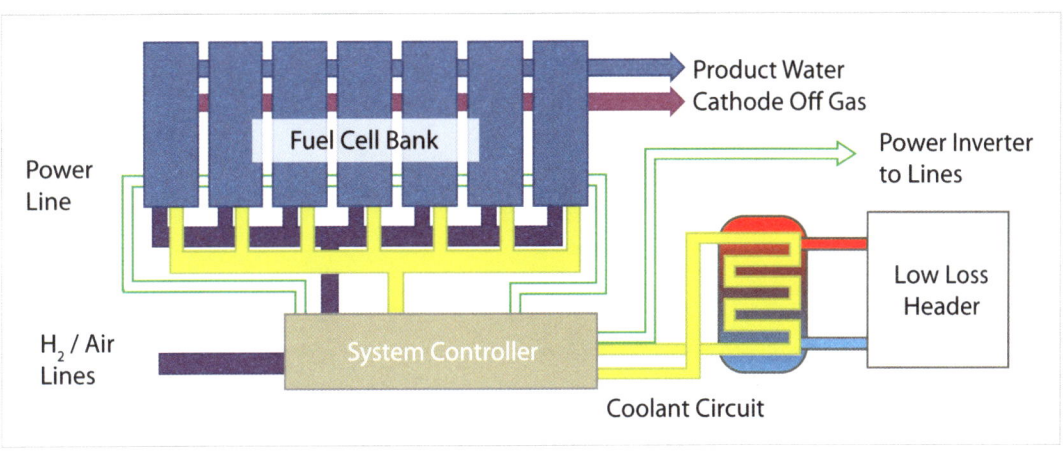

T   +44 (0)7790 517484
E   simon_gilbert@hotmail.com

T   +44(0)7793 556375
E   nataliekimmorris@msn.com

HUMAN POWERED POTABLE WATER PROJECT

MECHANICAL ENGINEERING MEng
MECHANICAL ENGINERING WITH BUILDING SERVICES MEng

Responses to Climate Change

Navid, James, Jonathan and Matthew, the members of this design group, have been studying together for a minimum of four years and this project has concluded their studies. All group members are following disciplines of Core Mechanical Engineering or Mechanical Engineering with Building Services. The team are now looking forward to starting their professional careers within the engineering field.

The flywheel assembly is the key to the system, trasnferring the energy supplied through the crank by the user into useful energy to heat and cool the water for purification.

T   +44 (0)7859 877402
E   naviderfan@hotmail.com

The crank, gear and chain system were extensively tested virtually using computer aided engineering techniques, and physically on a full scale prototype. The design was implemented to make the most efficient use of the energy.

This design is to be judged at the international ASME Student Design Competition 2007.

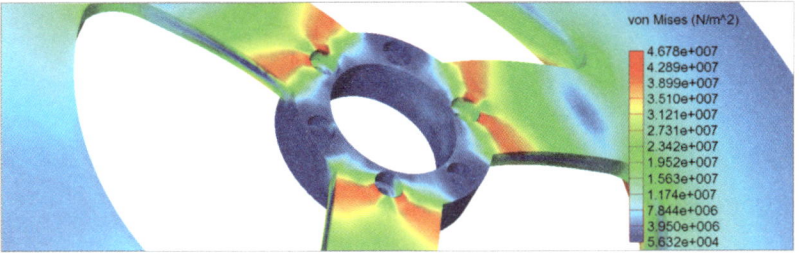

113

SOLAR POWERED REMOTE CONTROLLED MODEL AIRCRAFT

AEROSPACE ENGINEERING MEng
MECHANICAL ENGINEERING MEng
MECHANICAL ENGINEERING WITH AERONAUTICS MEng

Responses to Climate Change

Mgbechi Onuora, Benjamin Ozdemir, Pritesh Mistry, Pragnesh Patel

The aspiration of this project was to design, manufacture, and test a fully functional remote controlled model aircraft to investigate the feasibility of using solar energy as a fuel source in the application of aircraft propulsion.

This final year project was proposed to the department by the students as an environmentally friendly venture. All members of the group are currently looking forward to starting their professional careers in their respective chosen engineering disciplines.

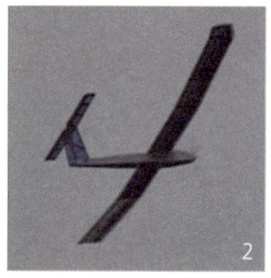

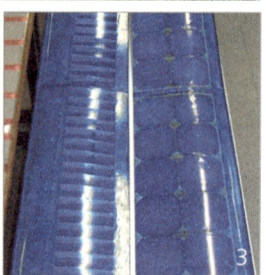

Mgbechi Onuora
T   +44 (0)7960 103184

Benjamin Ozdemir
T   +44 (0)7732 428213
E   benozdemir@hotmail.co.uk

Pritesh Mistry
T   +44 (0)7905 976297
E   p.mistry26@yahoo.co.uk

Pragnesh Patel
T   +44 (0)7947 506264
E   pragz_189@hotmail.com

1   The complete structural assembly of the final model aircraft design
2   The aircraft during its test flight
3   The Sun Power A-300 solar cells
4   The structure of the wing
5   The team with the model aircraft pilot, Robert Mahoney
6   The detached wing and fuselage assemblies
7,8 The complete structural CAD design
9   Internal circuitry of the aircraft

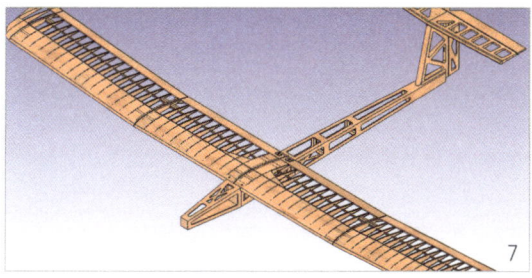

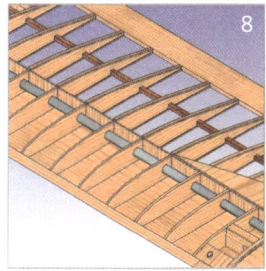

In a world where large amounts of fossil fuels are burnt in commercial aircraft flight it is necessary to find alternative fuel sources, which do not cause environment damage. This project explores one method of solving this problem.

In order to successfully meet the objectives of the project it was necessary to separate it into two separate aspects, solar power electrical circuitry and model aircraft design, even though both aspects of the project were interrelated.

The solar power part incorporates new technological advances in solar equipment. This included the utilisation of a Maximum Power Point Tracker (MPPT), which maximizes the available power of the system and the use of Sun Power A-300 solar cells (thin and flexible technology), which are recognised as the world's most efficient commercially available photovoltaic cells. The final circuit incorporates a buffer battery which stores excess solar energy and a speed controller for the brushed motor. Servo motors provide the direction control for aircraft.

The design specification stated that the aircraft should be lightweight and stable; hence the design was based on is a glider. Throughout the aircraft design process a number of computational engineering tools were employed. These include SolidWorks (3D CAD modelling), XFLR5 (aerodynamic analysis) and COSMOSWorks (stress analysis). These techniques were employed alongside conventional analytical methods which were used to determine the performance capabilities and final dimensions of the aircraft.

The aircraft was manufactured using traditional model aircraft building materials and techniques. A method of low temperature soldering was employed to form the solar array module, which was then encapsulated using clear laminating film to provide module protection and even distribution of wing loading.

A successful preliminary non-solar flight test was carried using battery power. This test effectively demonstrated pre-determined manoeuvres such as take-off (via hand launch), climb, turn, cruise and land. This clearly showed that a specially built powered glider designed to carry solar cells on the wing had satisfactory aerodynamic and structural characteristics. Although independent solar flight is yet to take place during the summer, design calculations and experimental results have suggested the aircraft to be capable of climbing at 9.74 m/s, which can be maintained for 4 minutes at an angle of attack of 3° after a 3 minute charge cycle.

WELLBEING, HEALTH AND DESIGN

Products designed to make us feel better add to our quality of life. New ideas in this field go further than ever before to enhance our physical and emotional wellbeing. Understanding people's needs and aspirations is the primary starting point of these creative industrialists.

HALIDU ABU-BAKAR

MULTIMEDIA TECHNOLOGY
AND DESIGN BSc

Wellbeing, Health and Design

I have acquired a variety of skills on the course but I am particularly keen on 3D design, Adobe Flash animation and designing interactive contents.

T   +44 (0)7852 348974
E   h.a@halidu.co.uk
W   www.halidu.co.uk

The Missing Link is a multimedia artefact that depicts a continuous ambient and scenic journey featuring elements of nature known to have a soothing and calming effect on the general state of mind. My inspiration derives from research into musical and visual therapy for autistic people.

Every single graphic, from moving objects to special visual effect are 100% computer generated, using 3ds Max (rendered in Mental Ray), After Effects and Photoshop. It features four scenes consisting of camera fly-through and animation against a tranquil musical background.

1 A set of design
guidelines was developed
for the aviation industry, to
help relieve the symptoms
of air travel stress and
claustrophobia. The
project was supported
by designers from
Virgin Atlantic Airways
and British Airways.

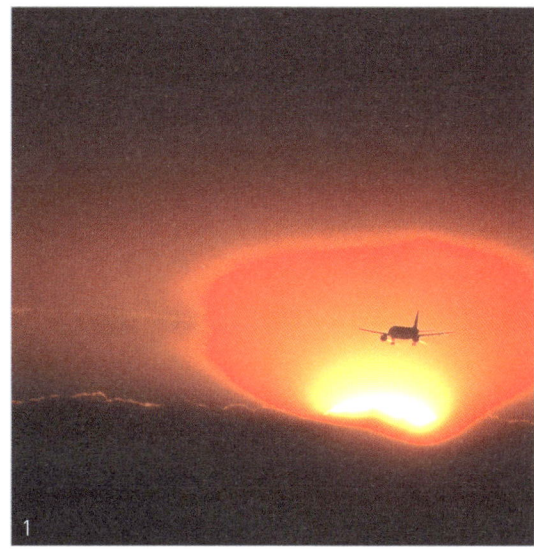

LAURA BAIRD

INDUSTRIAL DESIGN BSc

— Wellbeing, Health and Design —

I am particularly interested in
the concept of emotional design
– how the use of products and
environments can affect people's
moods. The power that this
bestows on designers has the
capacity to significantly increase
the psychological wellbeing of
consumers. Spending a year
teaching Design Technology has
given me the confidence and skills
to successfully manage projects
within a number of constraints
– and this knowledge was put
to good use as show team
coordinator for MADE IN BRUNEL.

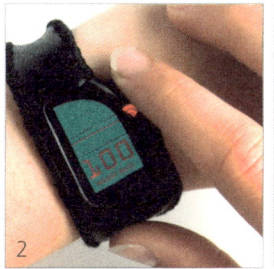

2 Conceptual design for The North Face brand. The
wrist band monitors various indicators of good health
through a sensor on the underside. 3 Enigmatic
lighting design, which reached the finals of the Lighting
Association's 'Student Lighting Design Awards' 2005.
The lamp reflects an image on the adjacent wall which
contradicts the image painted on the shade.

T   +44 (0)7786 542524
E   laurabaird@hotmail.com

DENISE BUTCHER

INDUSTRIAL DESIGN BSc

— Wellbeing, Health and Design —

During my placement year I worked for an advertising and marketing company assisting in website design.

In my final year, I worked alongside Kingsdown Special Needs School in creating a piece of equipment that would help the children understand the concept of cause and effect.

T   +44 (0)7708 743420
E   botchit@hotmail.com

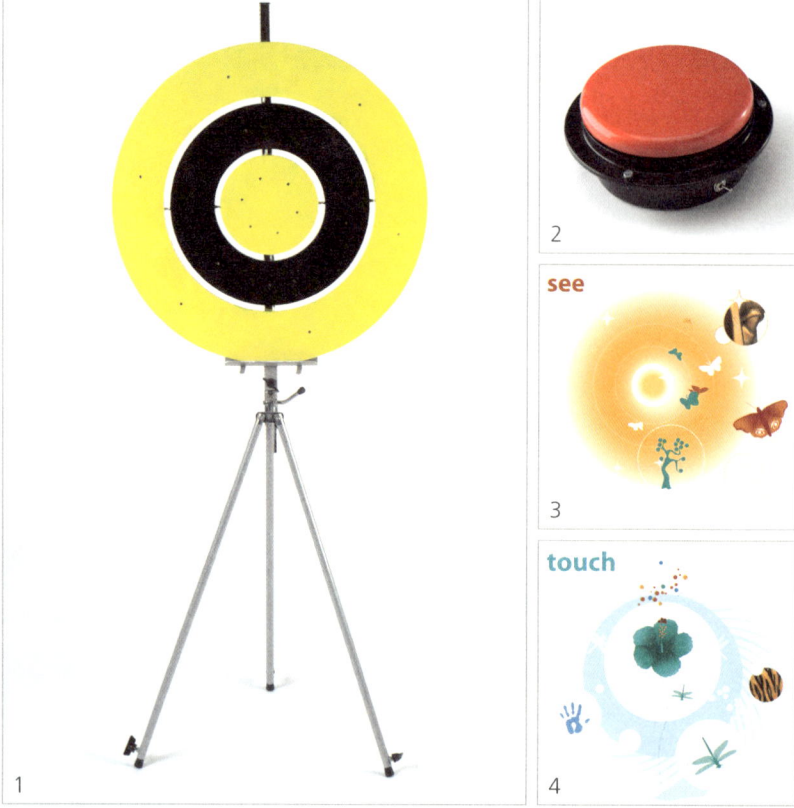

see

touch

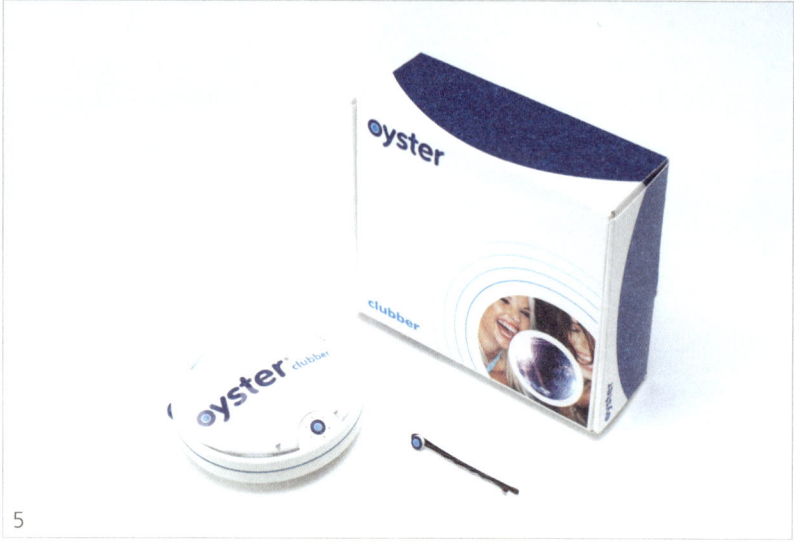

1 – 4 The target that was designed as part of my major project aimed to help children with special needs understand cause and effect. It consisted of an electronic circuit that contained a vibration sensor that would set off an alarm and light up when a ball made contact. This was extremely successful and enjoyed by the children. A switch was also created using radio frequency technology. The child would press the switch and the transmitter would send a signal over to the receiver that would trigger the solenoid and release a catapult. 5 Oyster Clubber - the ultimate clubbing accessory. It can download music played at a club, send messages, take photos, display your I.D. and be used as a payment method.

Feet Treat is a concept foot massager, designed to help maintain balance and reduce the risk of falls within certain, at-risk individuals. It helps strengthen the legs and maintains the sensomatory system whilst providing a pleasurable massage. The architecture encourages paired use, providing more incentive to be active; Feet Treat was derived through in-depth research and detailed user profiling, along with a variety of other suitable concepts.

JAY CANHAM

INDUSTRIAL DESIGN BSc

— Wellbeing, Health and Design —

For me, design does not mean 'Designer'. I see design as a process, a method and approach to problem solving, with infinite possibility.

"Design has become the most powerful tool with which man shapes his tools and environments (and by extension, society and himself)." Papanek

It makes me smile and keeps me thinking; the shear possibility it brings should excite us all.

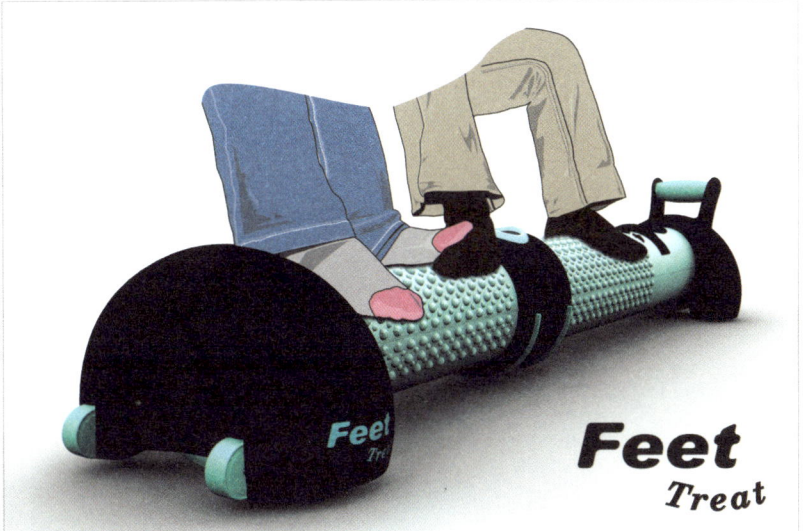

Breathe Deep is a fresh air diffuser designed as part of a product range for Original Source, intending to expand the brand into the health and wellbeing market sector. Twisting at the base opens the diffuser; releasing a chilled breath of fresh air to de-stress and invigorate.

T   +44 (0)7980 342505
E   jayceyned@hotmail.com

TERENCE DE'SOUZA

MECHANICAL ENGINEERING
AND DESIGN MEng

— Wellbeing, Health and Design —

My fascination with engineering and design has motivated me to fully develop my skills and fulfil my passion for both disciplines on the MEAD course. In the last 5 years I have developed fundamental engineering and creative thinking skills.

My placement year spent working in the architectural industry has given me a greater insight into other aspects of the design process. In developing my knowledge and understanding, Brunel has prepared me for the engineering and design industry.

T   +44 (0)7790 131690
E   terencedesouza@hotmail.com

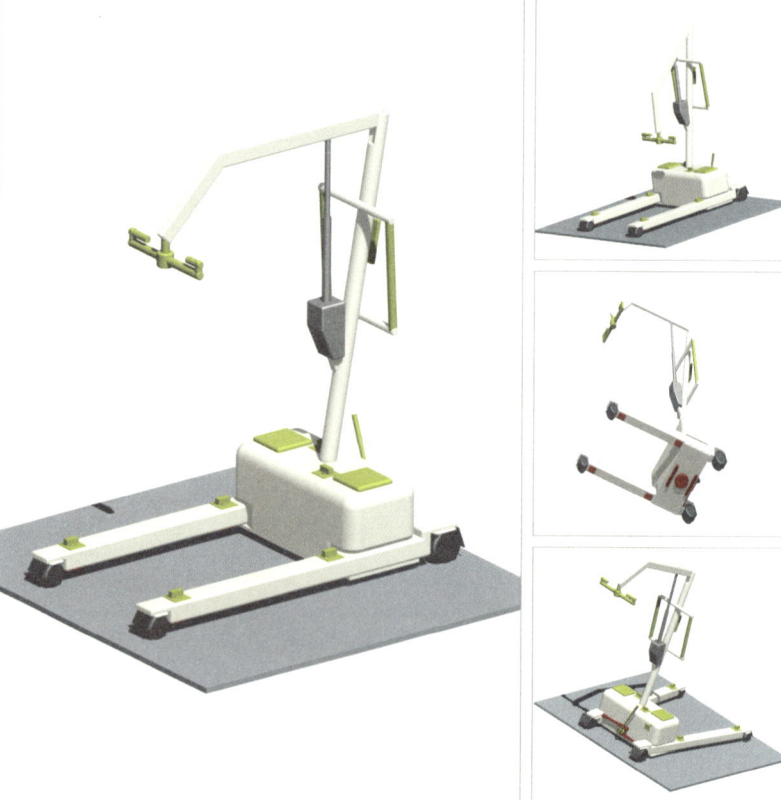

Hybrid hoist combines the functions of mobile and bathing hoists into one product. The hoist is safe to control and requires minimal modification to the bathroom.

The Chef Joe's Spice Grinder enables anyone to create exceptional dishes. The grinder comes in African, Asian and Indian spice selections. Simply select the dish from the grinder and dispense the aromatic spices.

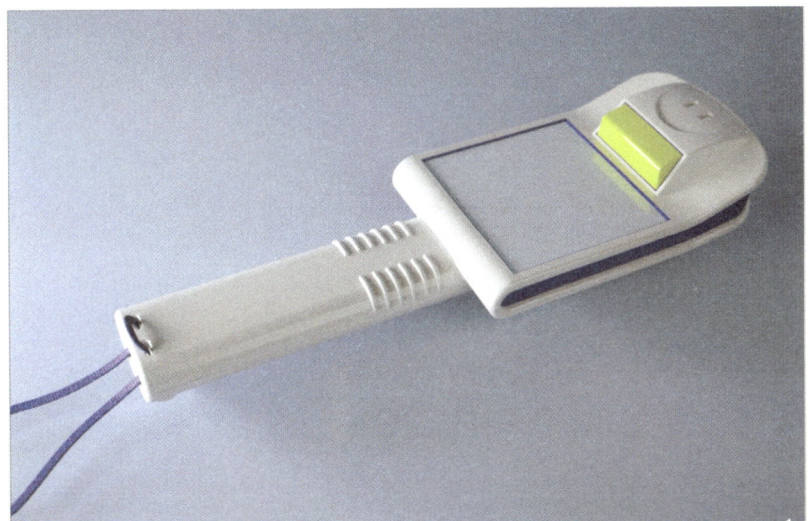

OWEN GLYNN

INDUSTRIAL DESIGN BSc

- Wellbeing, Health and Design -

1, 2 eSee evolved from a design exploration into the needs of visually impaired people. Intended to aid supermarket shopping, eSee uses Radio Frequency Identification to scan items. The intention was to create an accessible product for people with varying needs and abilities by incorporating a graphical and audible interface in a comfortable, intuitive product.

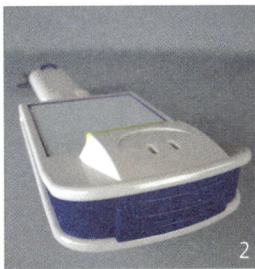

A year working in Shanghai gave me the opportunities to further my skills as a designer. I gained a new insight into the industry through working with Chinese factories and international clients, and developed a deeper appreciation of cultural diversity and its influence upon design. With this, I leave Brunel enthused and motivated, looking forward to becoming involved in the global design industry.

3 A future product direction for Durex aiming to integrate the core values of the brand into simple, elegant items which help to support the development of the relationship. 'Feel' identifies the difficulties and isolation felt by both partners during pregnancy and unifies the couple with thoughts and feelings during the day. 4 Examples from a pencil range designed for Shanghai Marco Stationery.

T   +44 (0)7791 411372
E   owen.glynn@gmail.com

MATTHEW HIGHAM

INDUSTRIAL DESIGN BSc

— Wellbeing, Health and Design —

The creativity, attention to detail and professionalism that I have developed over the past four years, which has included working with PDD and Hyphen Design, have enabled me to not only create products that look good and function well, but the experience has also allowed me to understand the wider impacts of design, on society and our environment. Design for me is entering an exciting stage, and I look forward to being able to apply and further develop my skills in this rapidly changing and flexible industry.

T   +44 (0)7890 434115
E   matthew@matthewhigham.co.uk
W   www.matthewhigham.co.uk

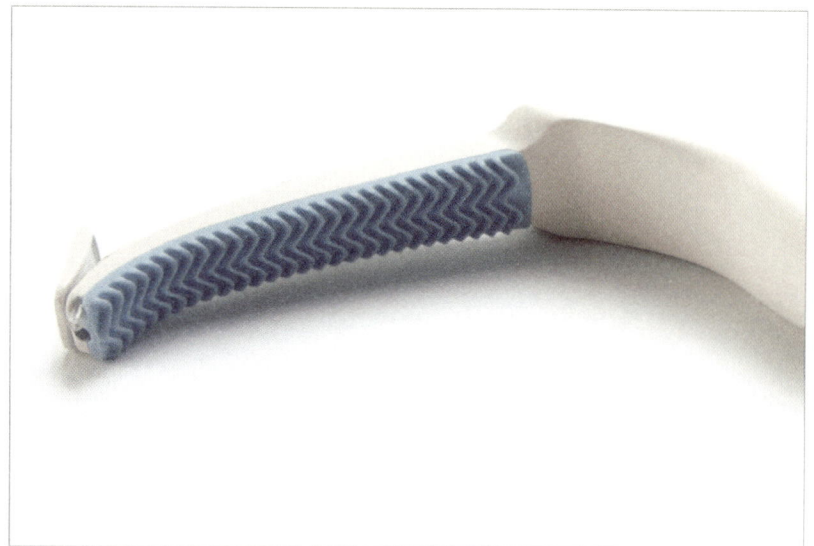

IntroBAL Tuberculosis is a worldwide epidemic, killing 5000 people everyday. IntroBAL has been developed in collaboration with NHS Innovations London to form part of a revolutionary TB diagnostic procedure. This simple, inexpensive device allows a doctor to visualise the vocal cords in order to introduce a catheter. Improvements brought about by the use of this diagnostic technology could dramatically reduce deaths from TB.

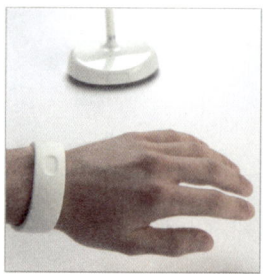

Connect A future product direction for Durex aiming to integrate the core values of the brand into simple, elegant items which help to support the development of the relationship. Connect is a next generation baby monitor, allowing parents to maintain an emotional link with their child by subtly communicating the child's level of activity, or distress, via a delicate glow on the parent's wrist band.

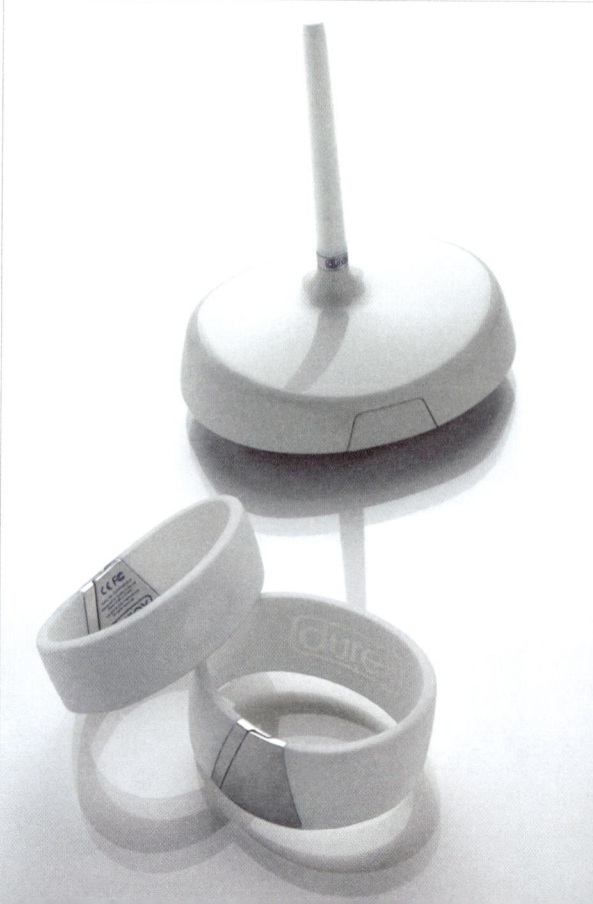

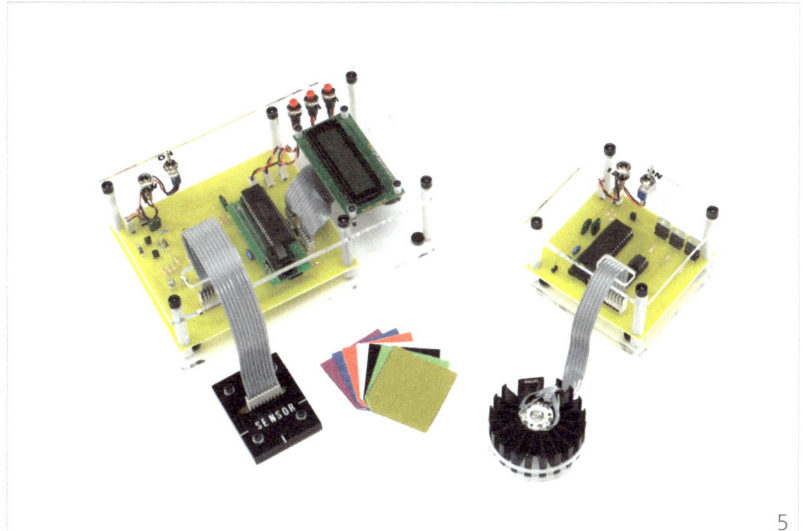

ALEXANDER HILL

INDUSTRIAL DESIGN
ENGINEERING BSc

— Wellbeing, Health and Design —

Since swapping the lab coat for the designer label and joining the Industrial Design Engineering course at Brunel, I have acquired the necessary skills to tackle design at all levels. I have utilised advanced engineering concepts, developed the detailed designs necessary to realise a product and explored form as well as function.

My previous experience working in medical research gives me a novel and mature approach coupled with direct experience of commercial design working at Aqualisa during my placement year.

1 – 4 Maxwell Render images based on Pro/ENGINEER models of a space-efficient, collapsible storage system to maintain patient samples in dry ice at -80°C.
5 Digital Adobe RGB colour scanner utilising colour space matrix transformations (left), pulsed width modulated driver for a Lamina RGB light engine (right).

T   +44 (0)7803 012314
E   dt03ash@hotmail.com

SOPHIA KELLY

INDUSTRIAL DESIGN BSc

The past few years have given me a series of diverse and interesting design projects covering all disciplines, providing me with the confidence to pursue a career in design.

A year at Metso Minerals introduced me to practical engineering while a further year at Yell Group gave me valuable experience in graphic design and working to tight deadlines. In addition, university life has allowed me to pursue the hobbies of climbing and set design for amateur dramatics.

T   +44 (0)7870 733014
E   sophiakelly@googlemail.com

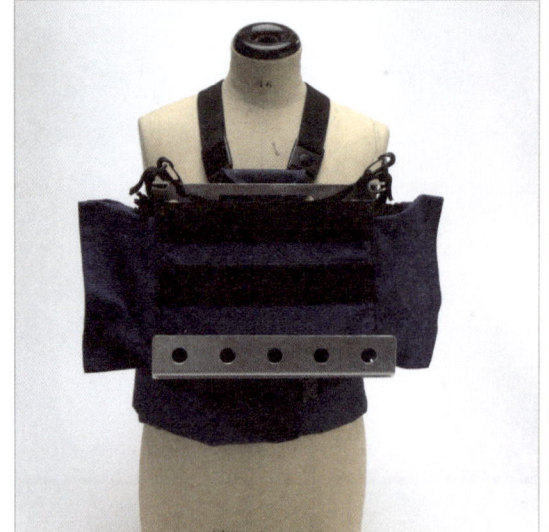

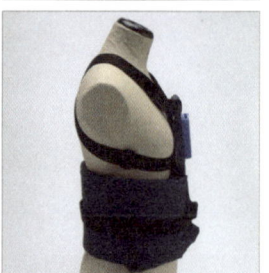

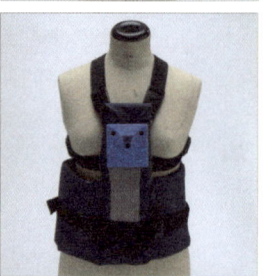

Uni-Clip Harness is an easy clipping system for attachment of a TV production soundman's bag. This bag weighs around 10kg and is carried for up to 18 hours a day. Designed to reduce the physical strain on the user's body, it is easier to use than the current models available.

Always Refresh combines an air freshener, light and hygiene spray making a public loo environment more pleasant.

SUNGMIN KIM

INDUSTRIAL DESIGN BSc

Wellbeing, Health and Design

The opportunity to explore my creativity and imagination has been central to the varied and fantastic experiences I have enjoyed at Brunel. I have a broad appreciation of design process and extensive design skills. As a result of my project work, I have become particularly interested in the fields of furniture design and housewares which are where I hope to focus future work as a designer. I will expect a smile on my face when my design comes out in the world.

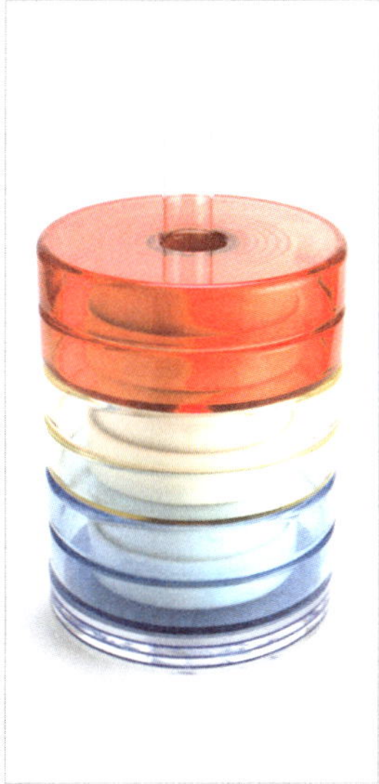

Jelly Stool Designed for children 3 to 10 years. Its height is adjustable, so that it accommodates children of different sizes. It is suitable for indoor and outdoor use. When folded, it can easily be stored in limited space such as a car boot. Children can use it as a stool, a table, or a toy. The Jelly Stool offers a combination of playful and functional design.

T   +44 (0)7903 717690
E   now2007@hotmail.com

LOUISE MCKILLOP

INDUSTRIAL DESIGN BSc

Wellbeing, Health and Design

A 12 month placement working within the plush development team at The Walt Disney Company has enabled me to return to my final year at Brunel with a professional, organised and creative approach to my projects. I have gained a great knowledge and understanding of the importance of designing for a specific user and meeting their requirements. I am aware of the challenge of designing products to meet the style constraints found in multinational companies and the necessity of intensive market research before beginning the design process.

T   +44 (0)7709 491480
E   louise_mckillop@hotmail.com

1 – 4 Winnie The Pooh Playtime is a developmental soft toy for children aged 6 to 12 months. By focusing on key sensory, cognitive, motor, emotional, social and language developments made by a child in a set age range I was able to develop a toy that helped enhance these milestones. Winnie The Pooh Playtime is currently being developed by The Walt Disney Company. 5 – 8 The North Face GPS Wristband is a contextual design for The North Face. This device would be equally at home on the slopes of Aspen or locating friends in London.

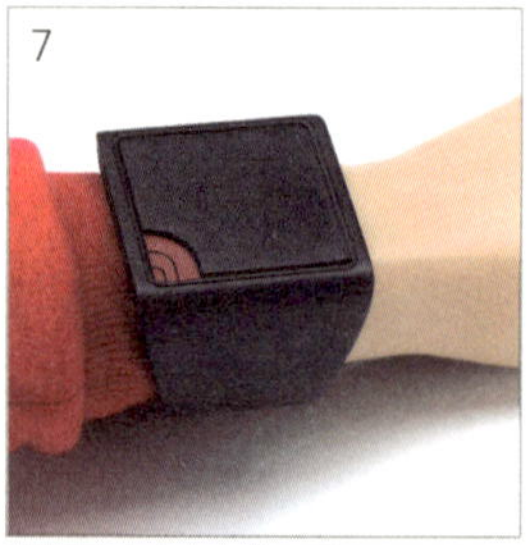

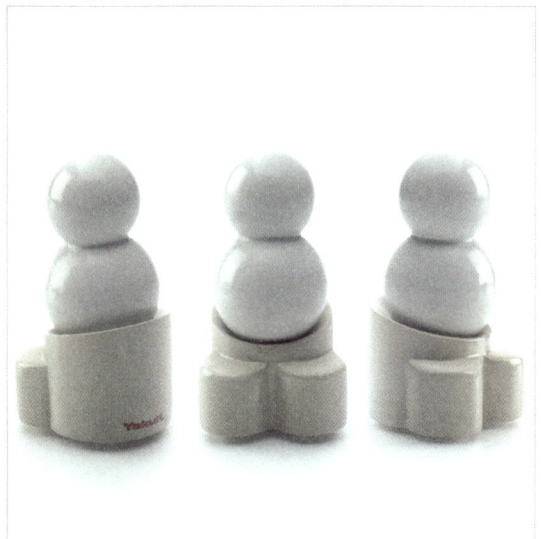

Yakult Tempo is a Little Yakult lights that pulsate a steady glow. Regulating your breathing to match the rhythm of the Yakult Tempo helps calm your body and mind, alleviating the build up of everyday stress.

QUANG NGUYEN

INDUSTRIAL DESIGN BSc

— Wellbeing, Health and Design —

My three years at Brunel along with my placement at Creation Studio Ltd have been very rewarding. I have made many friends as well as learning new programs such as Adobe Photoshop and Illustrator, Pro/ENGINEER, Microsoft Office, AutoCad, SolidWorks, Lightwave, Videotoaster and Adobe Premiere. In my spare time I like to play a wide range of sports and pride myself on my knowledge of films.

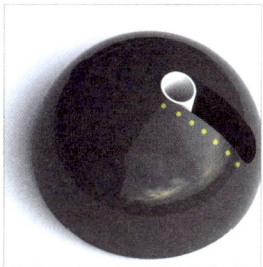

UniBowl is an adaptive bowling ball design which allows you to choose an appropriate finger size attachment to insert into the main body of any weight of bowling ball. The device measures the span of your hand to determine the most suitable attachment for your ball.

T   +44 (0)7754743868
E   quangojin@hotmail.com

LUKE PANNELL

INDUSTRIAL DESIGN BSc

Wellbeing, Health and Design

Over the past four years I have developed a wide array of skills ranging from concept design to structural analysis. I feel as comfortable working with my hands in the workshop as I am when in front of a computer or presenting to clients. I have a deep passion for the visually exciting and feel my strongest talents lie in product styling and graphic design.

My placement year at top office furniture design company Ahrend allowed me to further develop my 3D rendering skills and gain vital business experience producing work for companies including BMW, HSBC and EDF energy.

T   +44 (0)7881 908585
E   luke@lukepannell.com
W   www.lukepannell.com

Air Filtering Cycle Helmet A particle arresting powered respirator designed specifically for cyclists. It incorporates a dual stage filter that removes over 98% of all air-borne pollutants. In order to minimise the restrictions and discomfort placed on the cyclist, the product disregards conventional facemask design and instead incorporates a clear polycarbonate shield that makes no contact with the skin.

Future Visual Concept designed for the clothing brand Prada. The product incorporates wireless technology, flash memory and induction charging into a piece of unisex jewellery. Inspired by current vintage trends and Art Deco architecture the device comes with a selection of straps and can be worn as a necklace or attached to a belt or handbag.

1

2

3

CAROLINE PIERCEY

INDUSTRIAL DESIGN
AND TECHNOLOGY BA

— Wellbeing, Health and Design —

Kinderpendence is a portable toilet support for children with minor disabilities. Folding easily into a portable carry case, the toilet support can be carried by the child themselves. Kinderpendence was developed to give disabled children greater independence, a fundamental aspect of growing up for all children. **1, 2** Aesthetic prototype model. **3, 4** Working prototype images.

4

My time as a Brunel Design student has been an inspirational and enjoyable experience. Working at MERU designing bespoke products for disabled children greatly influenced my final year project. I find work in this field fulfilling and valuable. It has challenged me to be creative and innovative to maintain a strong focus on the end user's specific needs. My major project has inspired me to explore my design capabilities and further develop my passion for designing innovative products for specialist needs.

Sunshine is a concept product for Original Source. A clock designed specifically for the working hours of 9 to 5, the sphere emits a beam of light, mimicking the natural arc of the suns progression. It is a 'natural antidote to the stresses of life', ideal for people under pressure at work. The beam reaches the pinnacle at 5 o'clock, becoming warmer and brighter, symbolising the end of the working day.

T   +44 (0)7765 325974
E   cally_ally20@hotmail.com

STEPHANIE PRICHARD

INDUSTRIAL DESIGN BSc

Wellbeing, Health and Design

My time at Brunel has provided me with a firm grounding of the design process which will enable me to go forward in my future career.

Experience: 12 months as a designer at Reading Room, a leading digital communication company based in London.

Skills: Adobe Photoshop, Illustrator and InDesign, Pro/ENGINEER, Microsoft Office, sketching and modelmaking.

T   +44 (0)7877 175302
E   steph.prichard@gmail.com

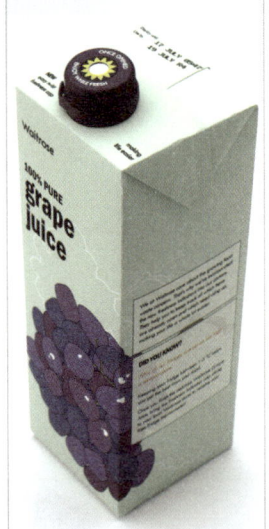

The Freshness Cap A simple and intuitive solution to minimising the amount of food and drink wasted by a household. It reminds people to use up their perishable items before they spoil by integrating time-temperature indicators into the existing packaging, which is automatically activated upon opening.

No Man is an Island An essay exploring the ecological potential of man-made islands to solve overpopulation and waste disposal in one fell swoop.

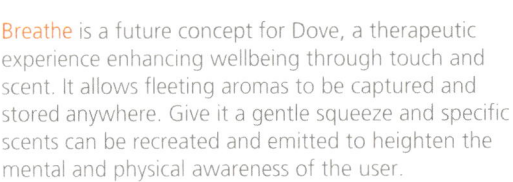

Breathe is a future concept for Dove, a therapeutic experience enhancing wellbeing through touch and scent. It allows fleeting aromas to be captured and stored anywhere. Give it a gentle squeeze and specific scents can be recreated and emitted to heighten the mental and physical awareness of the user.

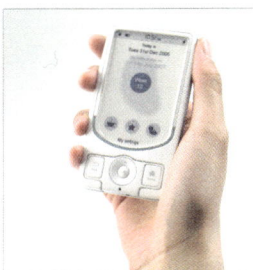

mamatoto An innovative, user-friendly, slimline portable device that reassuringly guides women through their pregnancy term. It provides an educational resource for first-time mums-to-be. The product provides weekly health tips, stages of the baby's development and monitoring the user's daily caffeine intake.

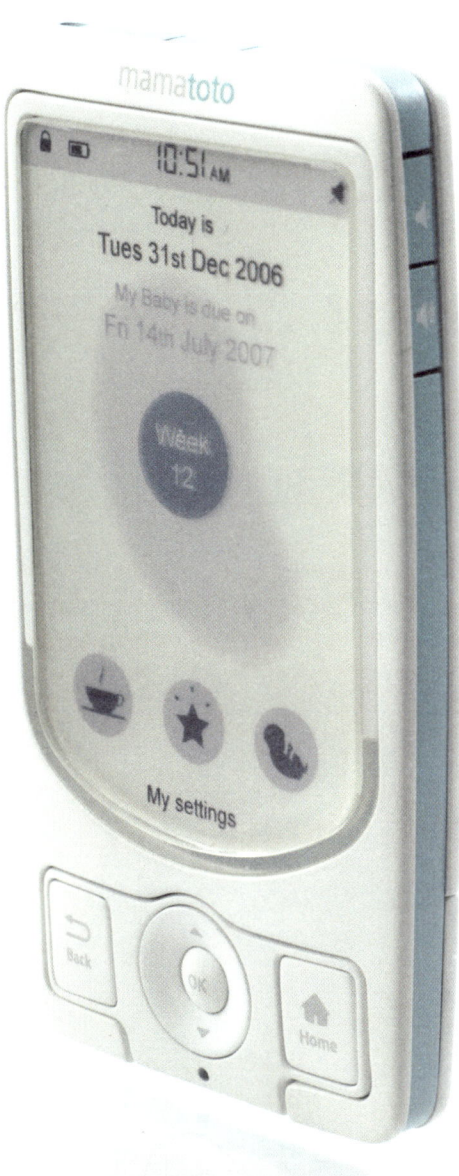

HEENA RAI

INDUSTRIAL DESIGN AND TECHNOLOGY BA

Wellbeing, Health and Design

In the essence of developing and designing products, I have worked to achieve a balanced relationship between function and aesthetics. My creative vision has been enhanced with an injection of enthusiasm, passion and focused energy. I am particularly keen on designing functional user-centric products and enjoy the challenge this brings. I am motivated by each stage of the design process from concept through to production. My experience of different cultures influences my design thoughts.

Experience: 12 months at Yell Group as a graphic designer creating adverts for many large clients including Volkswagen, Natwest and BMW.

Skills: Rhino, Adobe Photoshop, Illustrator, FreeHand and PIC programming as well as practical skills in modelmaking and sketching.

T +44 (0)7709 199818
E heenarai@hotmail.com

STRUAN RICKMAN

INDUSTRIAL DESIGN
AND TECHNOLOGY BA

— Wellbeing, Health and Design —

I have always enjoyed the process
of design and have found that
realising one's potential is one
of the most satisfying aspects.
There is always something
new to learn or a challenge
to overcome that continually
shapes my work. The time I spent
working and studying among
San Francisco's design community
has been one of the biggest
influences on my work.

1, 2 A public controller for a
Marmite game challenging lovers
and haters of Marmite to compete
against each other. 3 A syringe
sterilization kit for injecting drug
users to reduce the spread of HIV
and Hepatitis C. 4 A graphical
representation of harm reduction
as a method to combating HIV
and other blood-borne viruses.

T   +44 (0)7702 488973
E   struanis@hotmail.com
W   www.struanrickman.com

2

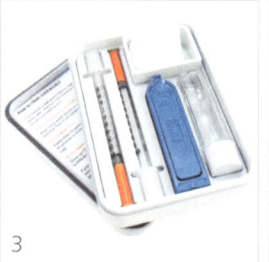

3

4

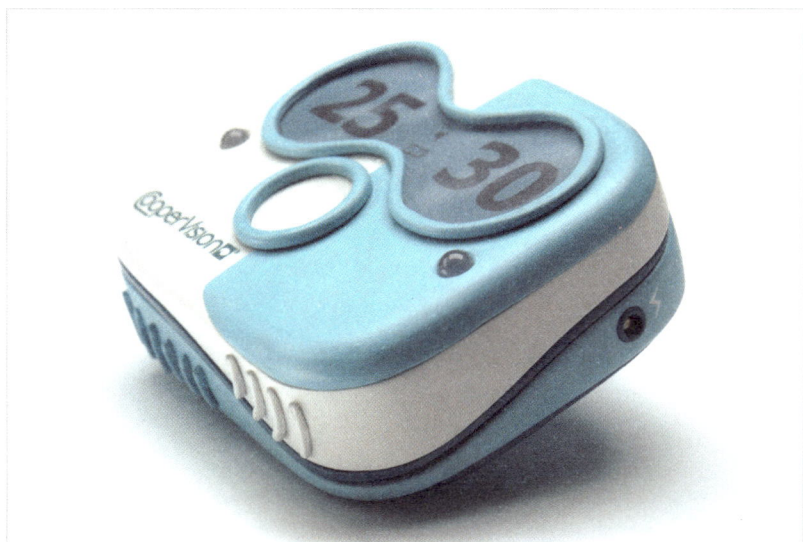

JANE SOWDEN

INDUSTRIAL DESIGN BSc

— Wellbeing, Health and Design —

iProtect Contact Lens Holder has been designed to improve the contact lens experience and to reduce the risk of infections. This has been achieved by studying the routine of contact lens users, experts and the market. The iProtect disposes of the contact lens solution each time the holder is opened. It also includes two thirty-day counters so the user knows when to replace the lenses the following month.

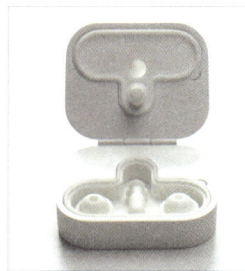

I am passionate about design and the experience I have gained at Brunel has been invaluable to achieving my goal – to become a successful and innovative designer. I hope to continue my career with the same drive and determination as throughout the last four years, using my design background as a stable base. I look forward to the future and am looking for my next challenge.

Original Source Bubble is a future concept for the 100% natural, vibrant and skin-tingling shower gel company, Original Source. The Original Source Bubble relaxes the worker during a stressful day by releasing bubbles that are 'packed with natural stuff'. Squeeze the globe to feel the tension drain from your body and smell the scent of fresh strawberries, lemon zest or forest pines.

T   +44 (0)7840 159736
E   email@janesowden.co.uk
W   www.janesowden.co.uk

LINDSEY STEVENSON

INDUSTRIAL DESIGN ENGINEERING BSc

— Wellbeing, Health and Design —

During my years studying design and engineering, including a year in industry spent at MKW Engineering, I have had the opportunity to use a number of approaches to develop engineered products and hone my design skills to a high standard. The products and projects that I have worked on have helped me to mature. Some of my designs are already on the market, and I hope my sound and wide-ranging foundation can only improve in the future.

T   +44 (0)7814 595511
E   linzi.stevenson@
    btopenworld.com
W   uk.geocities.com/linzi.steven-
    son@btinternet.com/

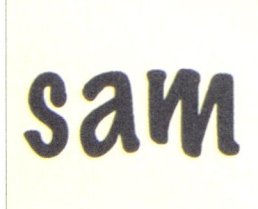

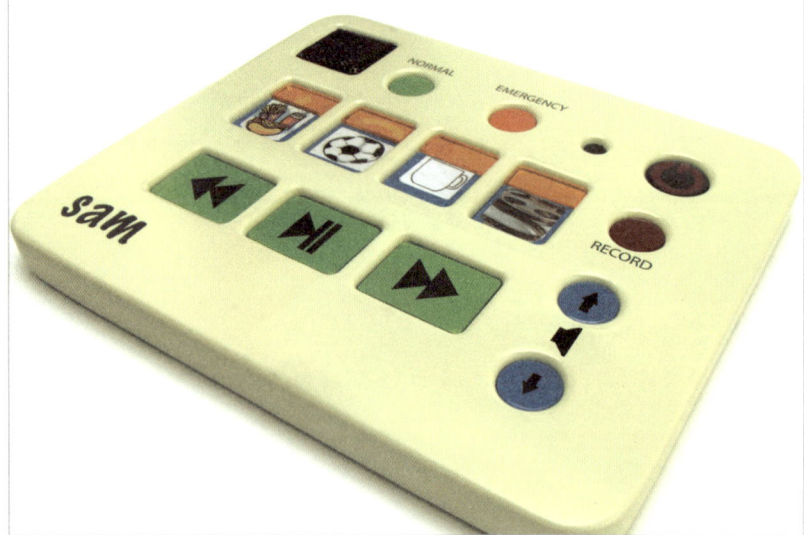

sam is a unique device for young adults suffering from Cerebral Palsy, who want music easily to hand. Primarily, this is a stylishly designed MP3 entertainment system for those with fine motor problems. It also holds four everyday messages, digitally stored and easily re-recorded, to help the user achieve a sense of independence. For emergency situations, the user can activate a set of key help messages from function buttons illuminated by changing the state of the device. A proving principle prototype and aesthetic model were created, as was an example of sam with a unique coloured skin modelled in Pro/ENGINEER.

Circus Top Hat is a toy for toddlers to help them to learn colours and shapes. Inserting a shape turns on the coloured LED on the hole and causes the shape name and colour to be spoken out loud and displayed on the screen. Multilingual flower displays the temperature in English, French and German at the touch of a button. Once the temperature reaches over 25°C the fan turns on to cool the user.

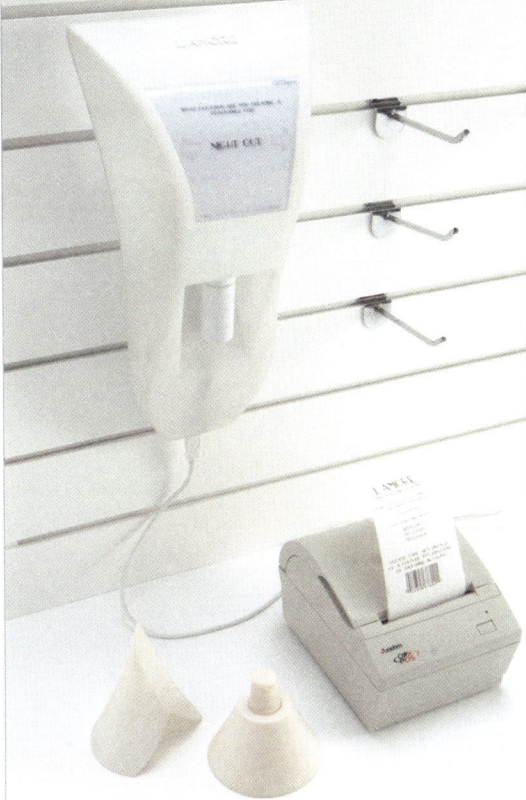

CHRIS TARLING

INDUSTRIAL DESIGN BSc

— Wellbeing, Health and Design —

Innovation, forward thinking and a professional approach to work are some of the key skills I will take from my time spent at Brunel. Runnymede has inspired the designer within and work experience for OTO3D and Procter and Gamble as both a modelmaker and Technical Designer, has given me maturity and a belief in my highly qualified skills.

Family, friends and my girlfriend provided support as well as strength and laughter, but above all memories to last a lifetime. Skills include proficiency in SolidWorks, Adobe Illustrator and Photoshop, and a Hesta blow moulding qualification. I am also a Brunel Student Ambassador and have completed the Duke of Edinburgh Award Scheme.

T   +44 (0)7739 314846
E   chris@tarlings.co.uk

L'amore Personalising fragrance unit that allows women to accessorise their outfit by way of scent creation. L'amore empowers female consumers fascination with fragrance, identity, fashion and beauty.
Water Storage Aid A Red Cross bottled water cap that encourages consumers to complete the equation of donation and relief, using water conservation as the key issue.

NATALIE VANNS

INDUSTRIAL DESIGN
AND TECHNOLOGY BA

— Wellbeing, Health and Design —

I have a keen interest in
user-centred and inclusive design.
Successful design harmonises an
attractive form with the functional
elements of a product. Subtle
changes to form and interface can
alter how a product is understood
by the user. This eye for detail
and understanding of interface
clarity were developed during
my year's employment at Nokia.

Skills: CATIA, modelmaking,
liaison with external
suppliers, Adobe Photoshop,
Illustrator and InDesign.

T   +44 (0)7941 828239
E   natalie.vanns@
    googlemail.com

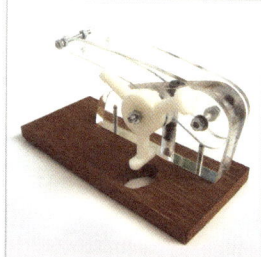

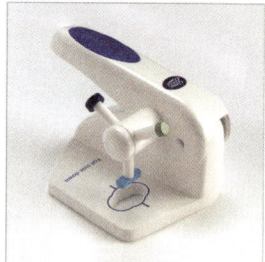

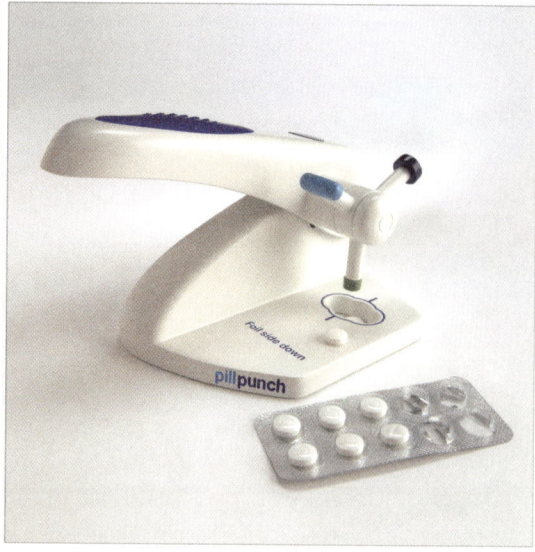

Pillpunch aids MS sufferers in the daily task of
dispensing blister pack medication. The product
uses mechanical advantage to accommodate for
the user's reduced strength. Multiple punch heads
ensure a range of pills can be dispensed. The model
above is a fully functional alpha prototype.

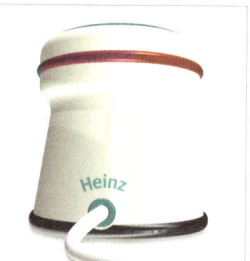

Pioneer commissioned my design to represent
their company as the sponsors of the BAFTA
Television Awards. The Pioneer Audience
Award has been used for the past three years
and has achieved great media coverage.
Touch Point is a conceptual product for Heinz.
Light is used for silent communication between
close friends and relatives, letting them know
you are safe and well, and thinking of them.

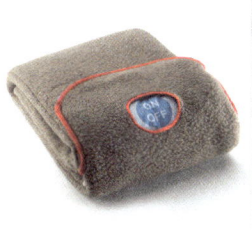

Cosy All the Time (CAT) is an elderly heating system, designed to tackle the 29,000 deaths that occur each winter within the UK's elderly population. The final outcome aimed to be as inclusive as possible embodying a user-centred design approach. Developed in collaboration with pioneering healthcare leaders Lightweight Medical and Inditherm plc. The project will be published in the 2007 DBAD Annual.

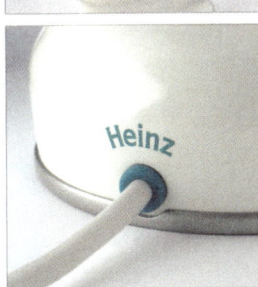

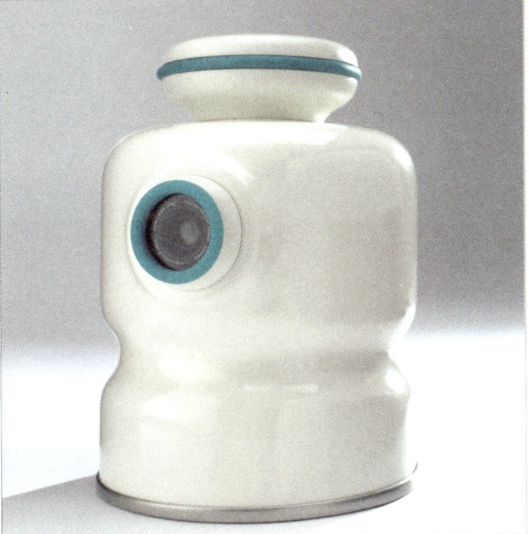

Heinz Magic Merit Board is a future concept for Heinz, reassuring parents that their children are getting good markz everyday. A star at school is stamped onto an interactive projection and is reflected at home in a selection of star shaped fridge magnets.

SAMUEL WELLER

INDUSTRIAL DESIGN BSc

- Wellbeing, Health and Design -

Four years at Brunel has culminated in understanding that successful design revolves around addressing real human needs, practising sustainable development and ensuring financial return. Good design is design that makes sense at all of these levels.

I enjoy both surfing the oceans and snowboarding the mountains of the world.

Experience: 6 months San Francisco State University 9 months Industrial Design at Huhtamaki Ltd.

T   +44 (0)7751 090622
E   sammyj85@fsmail.net
W   www.samwellerdesign.com

TSINGHUA UNIVERSITY
清華大學

The Academy of Arts and Design at Tsinghua University continually seeks to expand the students' horizon to improve their comprehensive quality, whilst strengthening the teaching in fundamental priniciples.

TANG ZHENQI
汤震启

INDUSTRIAL DESIGN BSc

Tsinghua University

I have acquired a solid grounding in design, both in theory and in practice. I have participated in several international design competitions, and been involved in product design products and design research for Nokia and Panasonic.

在过去的4年中我执着于设计，并习得基本的设计理念与技法。同时本人还参加许多国内外设计竞赛，并参与为诺基亚、松下以及卫浴公司所做的产品设计及设计研究等项目。

The design of a more updated computer model. To save resources, the design allows office or family members to each have their own on-demand terminals.

本设计提出一种较以往更新的电脑操作模式。出于节省资源目的，在我的设计 中允许家庭或办公室中每个成员按需拥有自己的终端。

E   tangzq@163.com
W   blog.sina.com.cn/tangzq

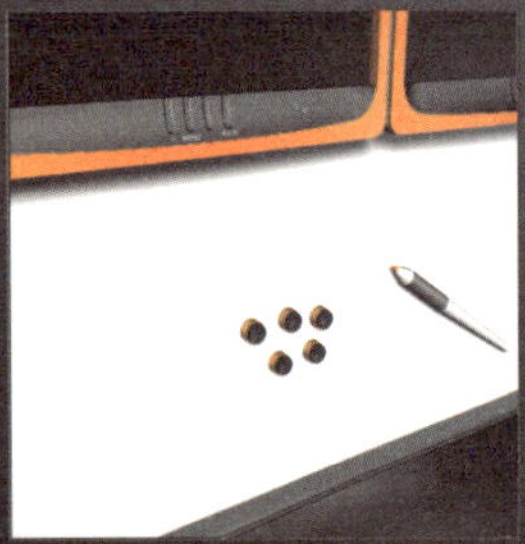

Satellite-Star Computer System

MENU   EXIT

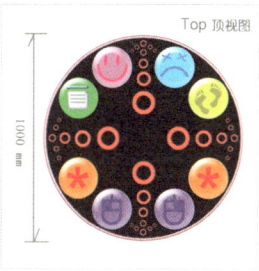
Top 顶视图

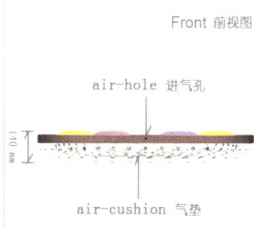
Front 前视图

air-hole 进气孔

air-cushion 气垫

Bottom 底面

transparent silica gel
透明硅胶

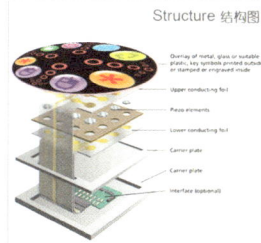
Structure 结构图

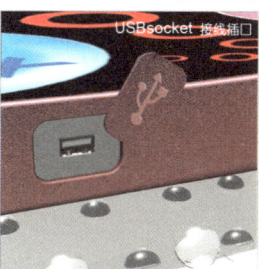
USBsocket 接线插口

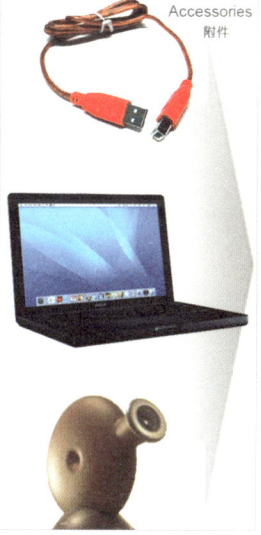
Accessories
附件

MIAO WANG
王淼

DESIGN STRATEGY AND
MANAGEMENT MA

- Tsinghua University

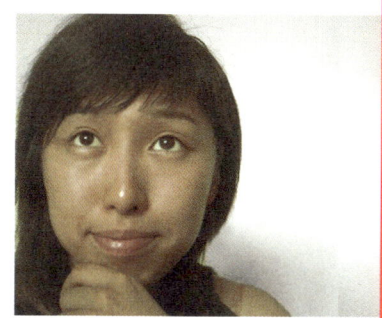

I'm a person of high responsibility, extrovert character and positive attitude, who does well in brainstorming and analytical skills, as well as sharing happiness during cooperative stages of design.

我是一个有责任心、热情开朗、积极上进的人，在团队合作中我擅长信息分析和概念创想工作，并乐于与队友们分享合作过程中的难得经验和快乐体验。

E   sunny08wm@gmail.com

The Personal Healthy Keyboard is a low technology product, with which office workers or long-time game players are encouraged to change the way they interface with the computer by operating with feet stamping instead of fingers pressing.

边走边跳: 个人健身键盘毯 以长期使用电脑的职员和沉溺于电脑游戏的青少年为目标用户群，通过阻性薄膜开关面板这一成熟的低技术，实现以脚部操作键盘进行网页浏览、音乐游戏、收发邮件等较简单的电脑操作内容，在休息手腕、眼睛的同时达到锻炼身体的作用。

ZHEN XU
許珍

INDUSTRIAL DESIGN BSc

Tsinghua University

I love design,and to me design is all about life and needs. I enjoy communicating with different people to make to a project a comprehensive success. A fresh and open mind is important to innovative design.

我是清华大学研一的学生.我热爱设计,对我来说设计是生活设计是人们的需求.做任何设计之前,我都会和不同的人交流获取设计灵感,对设计来说灵活的思维是重要的.

E   xu-z06@mails.tsinghua.edu.cn

Musical Dumbbell Ring targeted at those who are lack of exercise such as office workers and students.The combination of music and sports can offer people a chance to do exercise while listening to music.

哑铃-该设计针对目前大多数上班族或在校学生缺乏体育运动而设计的一款音乐哑铃。音乐与运动的结合既可以为人们提供听音乐锻炼身体的机会，同时也带来立体音效不绝于耳的震撼！

The MP3 player component of the dumbbell can be detached for easy recharging.

哑铃上的MP3可以取下来,然后充电.

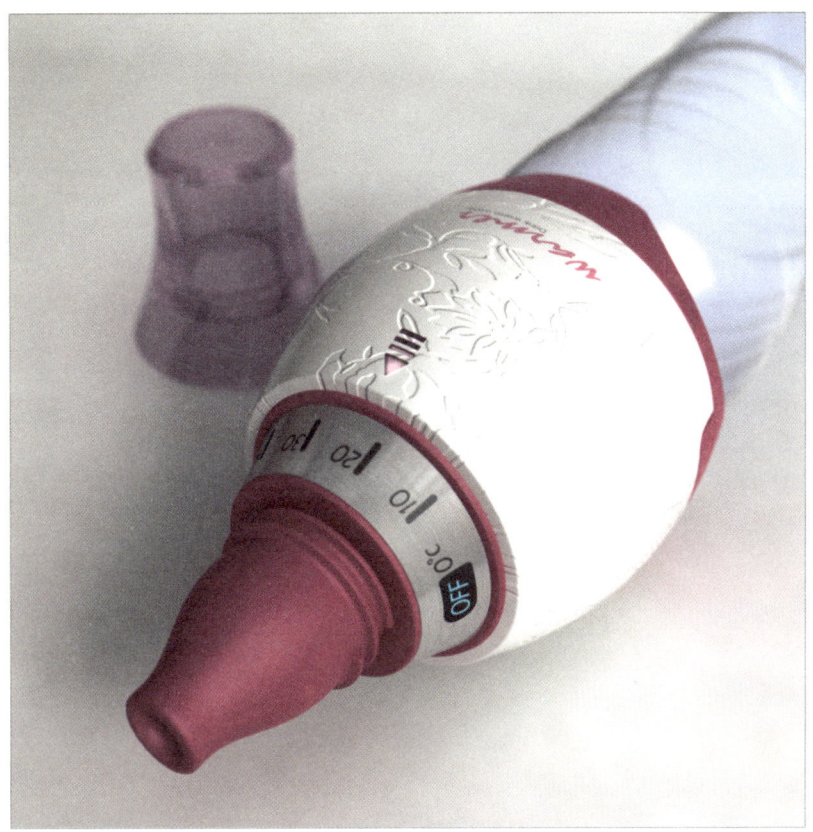

LI NAN
李楠

DESIGN STRATEGY AND
MANAGEMENT MA

- Tsinghua University

Throughout the course of my
studies so far, I have focused
upon focus the marketing
in relation design. I'm adept
in doing marketing strategy
analysis and developing
product marketing plans.

专业方向为设计战略和设计管
理，并对市场方面有所侧重。擅
长产品市场战略分析以及产品市
场推广策划。

Water warmer is an accessory of mineral water bottle.
By the rechargeable batteries' energy, it can warm up
the water which passes through. The temperature of
the water can be adjusted by user.

水加热器是一个与矿泉水瓶同时使用的附加物。通过
充电电池的能量来源，它可以将流过的水进行加热。
水温可以由使用者根据需要调节。

Most Chinese favour drinking warm water, especially
the elderly. Now they can drink warm water
conveniently even in winter. With this product, all they
need to do is to buy a bottle of mineral water.

多数的中国人喜欢饮用热水，尤其是老年人。现在本
产品可以方便的为其提供热水，甚至是在冬天，而他
们所需要做的只是去买一瓶矿泉水。

E   xiaoshi718@hotmail.com

ZHU WENJUN
朱文俊

INDUSTRIAL DESIGN BSc

Tsinghua University ———

With design and design management work experience, I am familiar with the product development process, as well as product processing technology, manufacturing techniques and I have an understanding of design from a commercial perspective.

有设计师及设计主管的工作经验。熟悉企业产品开发流程和产品加工工艺，能够从企业整体角度把握设计。

Comstuff can help you find items lost somewhere in the home. The stuff you want will beep and flash when you simply click the key on the base.

Comstuff 能够帮助我们找到一些忘在房屋某处的物品。只要简单的按一下主体上的按钮，您想要找的东西就会发出"嘀嘀"响声。

E   uniondesign@msn.com

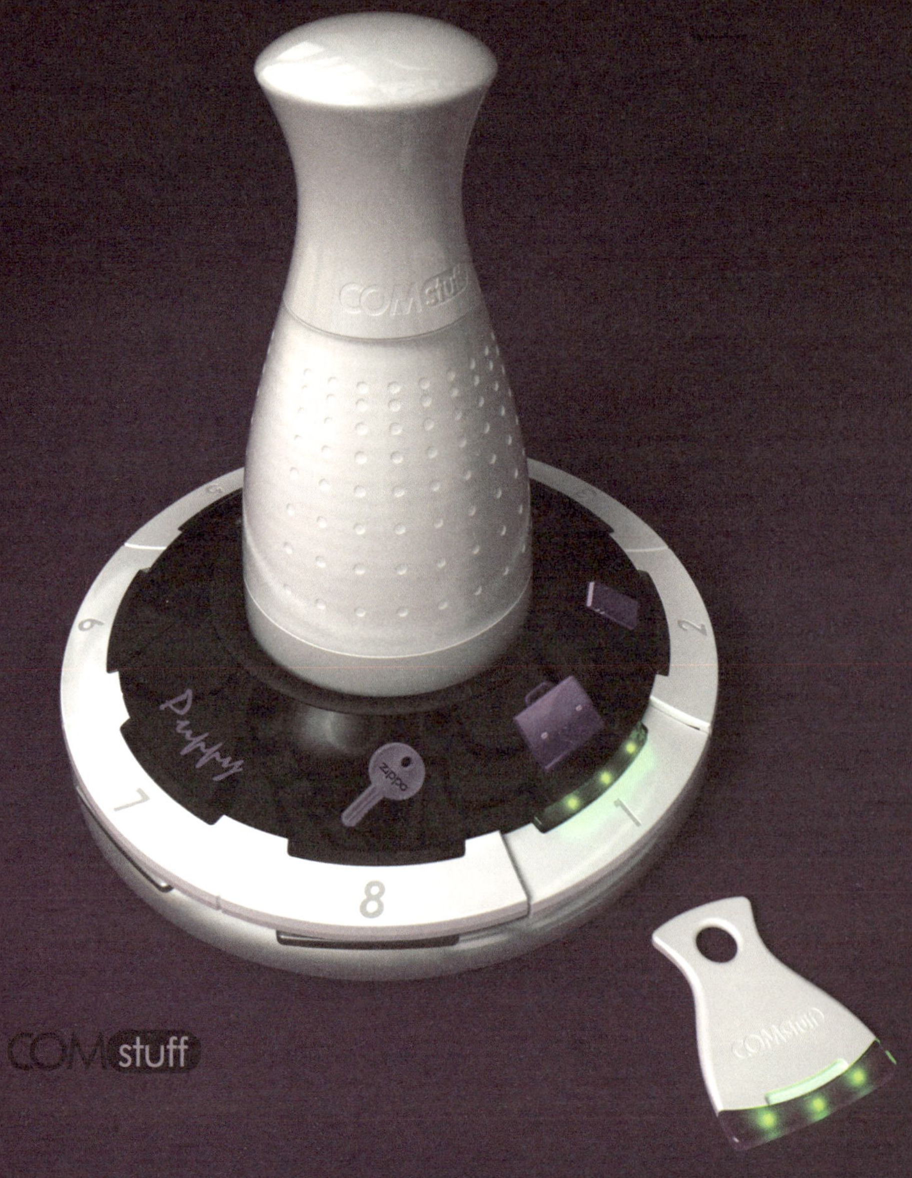

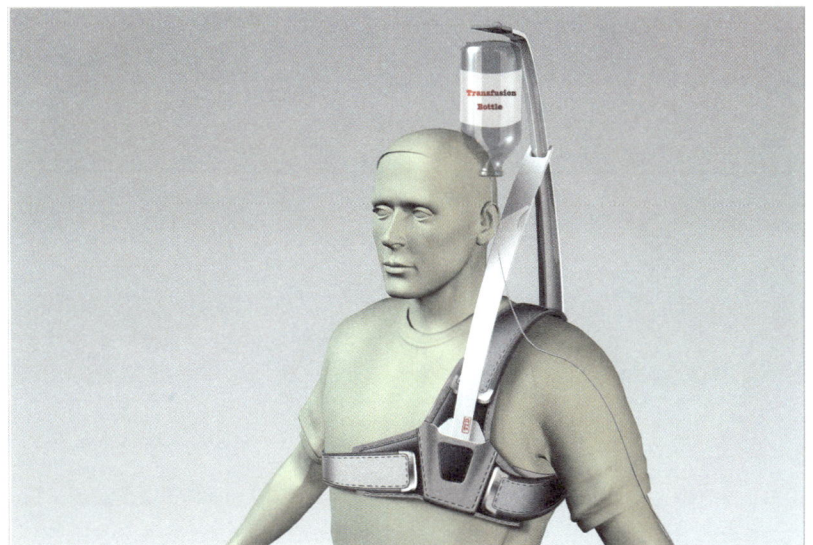

In China, there are many patients but few hospitals, so transfusion rooms are always overcrowded; particularly at times of flu prevalence. Existing transfusion equipment is difficult to move, which crowds patients into a small space, which is not conducive to recovery.

中国医院少, 人口多, 输液室人满为患, 尤其流感发生时。现有输液设备不便移动, 使患者集中在一个小空间, 心理感到压抑, 不利身体康复。

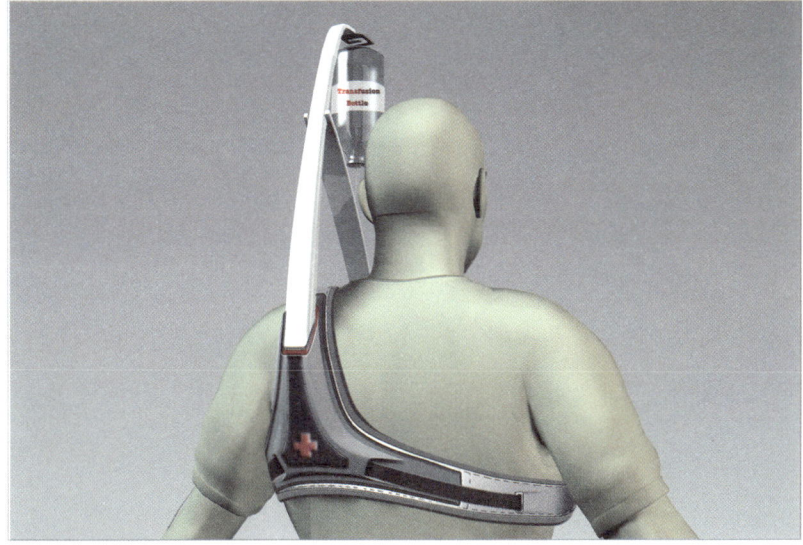

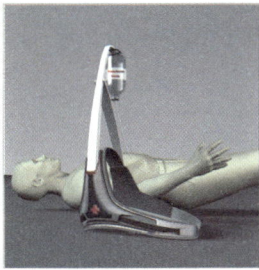

The design provides a transfusion backpack which is easy to move. It can offer patient fluid movement and support infusion bottles independently for patients who want to lie in bed.

该设计为输液者提供了一个便于移动的输液背包。输液背包可以独立支撑输液瓶, 以备患者卧床使用。

YANG JINCHAO
楊金潮

INDUSTRIAL DESIGN BSc

- **Tsinghua University**

Experience: Red Dot Design Concept Winner 2006; worked as an intern for a Telephone Products Designer in HAIER, 2007.

Skills: Certified in 3D/2D design software and product sketching, design management, colour and material design, and Chinese cultural studies.

E   sheavidhui@yahoo.com.cn

147

XIA MANG
夏芒

INDUSTRIAL DESIGN BSc

Tsinghua University

Experience: Worked as an intern of a Telephone Products Designer in Foxcon, 2005

Skills: Certified in 3D/2D Software and product sketching, design management, color and material design, comprehend Chinese culture

E   xiaminliang@126.com

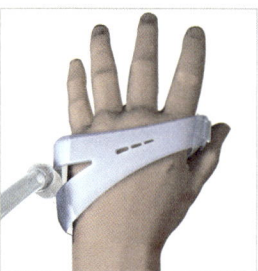

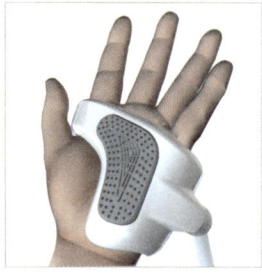

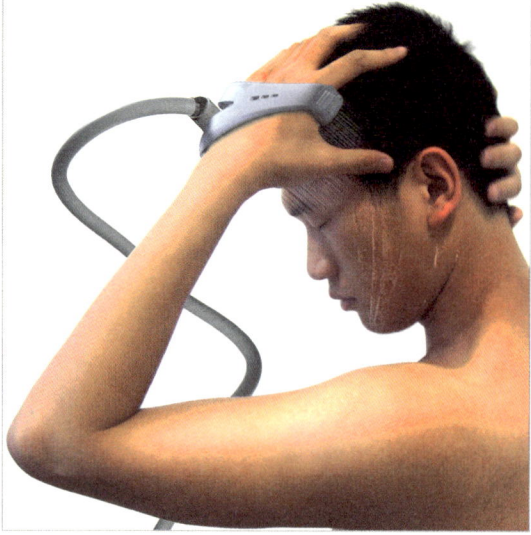

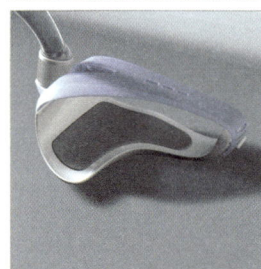

Existing shower heads waste water and are uncomfortable. We have to adjust the height many times during showering. 'Hand-shower' uses glove-styling, integrates hand and shower, and makes the shower process continuous, natural, and intuitive. 使用现有花洒需要频繁取放，浪费水并且不舒服。 'Hand-shower' 使用手套的造型，整合手和花洒，使得整个洗澡的过程连续且自然，让人感觉不到花洒的存在。)

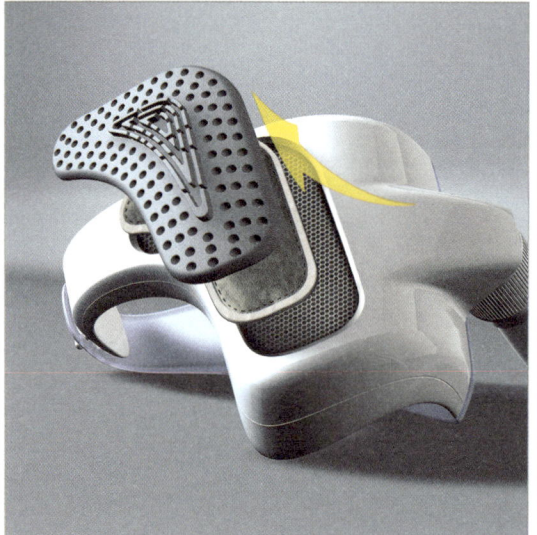

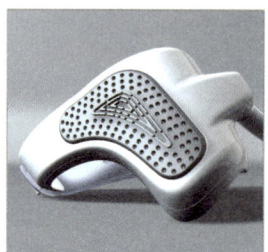

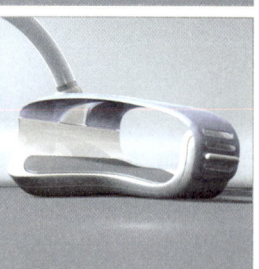

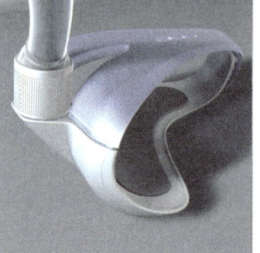

In addition, a packet of Chinese medicine can be put into the inside of it. Water flows through the dilution of Chinese medicine and sprayed on the body. It can clean the body and cure the skin.

另外， 'Hand-shower' 内胆里面可放置中药，水流可将其稀释并喷在身上，保养皮肤。

ACADEMY OF ARTS AND DESIGN, TSINGHUA UNIVERSITY BEIJING, CHINA
美術學院 ,清華大學 ,中國北京

At the Academy, great attention is paid to the studies of fine heritage of folk arts of all nationalities both at home and abroad. Also, attaching great importance to academic exchanges, the Academy keeps introducing and studying the latest foreign thoughts and methodology of fine arts and artistic design. It advocates the spirit of being precise in pursuing studies, relating theory to practice and seeking truth from fact, stresses the need to suit design to lives, the combination of design and craft manufacture and that of art and science. Students are trained to acutely grasp the changing trend and pace of social and cultural life. The Academy strives to create a lively academic atmosphere and fine educational environment.

At present, the Academy comprises three sections: Design, Fine Arts & History, along with the Research Institute and the School of Adult Education. The Design section consists of Departments in Textile & Fashion Design, Ceramic Design, Industrial Design, Environmental Art Design, Graphic Design and Information Art & Design. The Fine Arts section is made up of Departments in Arts & Crafts, Painting, Sculpture and Basic Teaching & Research Group. The Section of Arts & History is comprised of the Department of Arts & History and Editorial Office of the Arts & Design Magazine.

The Academy has a pool of celebrated art educators, artists and scholars. Currently among our 194 teachers, 51 are professors (18 advisers for doctoral candidates) and 84 associate professors. It boasts a strong faculty, complete teaching facilities and advanced experimental means.

Department of Industrial Design

The Department of Industrial Art was set up in 1975 and later became the Department of Industrial Design, one of the first founded in China. The Department of Industrial Design produces creative design professionals engaged in Product, Display and Vehicle Design for enterprises, research and educational organizations. The Department of Industrial Design currently offers three specialties: Product Design, Display Design and Vehicle Design.

Department of Information Art & Design

The Department of Information Art and Design advocates leading, interdisciplinary and high standards with open and internationalized thoughts in the running of the school. The aim; to bring up comprehensive professionals with concept's of quality on humanities and arts technologies, in addition developing students' abilities in art, science, creativity, integration and planning by the viewpoint of humanity.

INDIAN INSTITUTE OF TECHNOLOGY MADRAS

A faculty of international repute, a brilliant student community, excellent texhnical and support staff have all contributed to our pre-eminent status.

APURV SHRIVASTAVA AND VINEET BHANDARI

IIT Madras

Apurv Shrivastava
I am passionate about broadening my knowledge of all aspects of design. I would like to develop products from the initial concepts and ideas right through to 'design for manufacture'. Also I cannot help but mention my love for music. Rock, Pop, Jazz, Classical, traditional.. give me any genre of music, I love them all.

Vineet Bhandari
I like to take challenges in life with my key strategy as beat the deadline. I have a keen interest in product innovations. With projects like these, I gained experience encompassing every aspect of design and confidence to effectively communicate my thinking. In future, I would like to work towards solving problems faced in a global perspective.

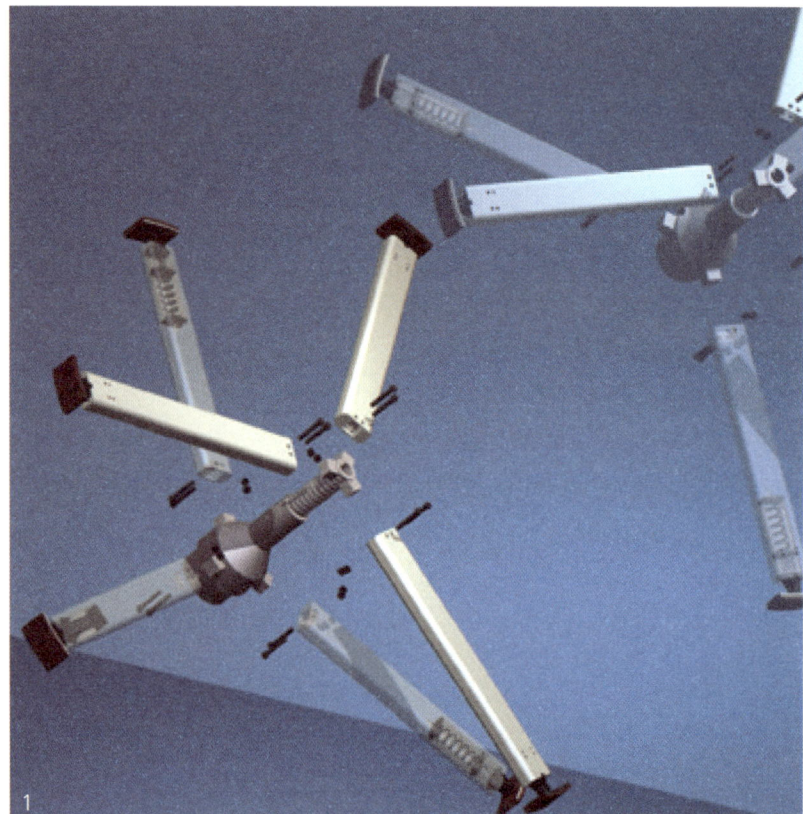

The objective of the project was to design equipment which will grind the inner surface of large cylindrical tankers (used in the shipping industry) to remove scale formed while carrying oil/chemicals. Currently, this work is done manually in harsh working conditions and takes a minimum time of two to three days to clean a single tanker.

We propose an equipment which has two sets of three grinding heads moving in a helical path grinding away the scales. One set performs the roughing action and the other set finishes the surface. The machine has adjustments to maintain a proper depth of cut, and can be programmed to run starting from one end of the tanker to the other end. We hope that it will clean the tanker in a period of few hours.

1 Exploded view. 2 Grinder head. 3 Our design modelled in Pro/ENGINEER. 4 A cut section of the tanker depicting the model. The image was generated by rendering the model in Pro/ENGINEER.

Apurv Shrivastava
E apurv4u2@gmail.com
T +91 9840582417

Vineet Bhandari
E vineet.iitm@gmail.com
T +91 9840131139

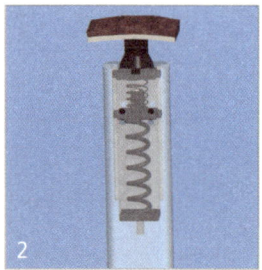

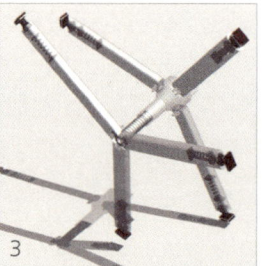

1

One of the main steps involved in bypass surgery is to suture the blood vessel with the aorta. To achieve this, a hole needs to be made on the surface of the aorta and the blood vessel needs to be sutured by placing it on this hole. The conventional method is to clamp the aorta but the clamp may slip during the operation and also may block the blood flow in the aorta.

We propose an alternate design where the blood vessel is first sutured to the aorta. Our instrument will then drill the hole on the surface of the aorta where the blood vessel has been sutured.

1 Sectioned view of the Instrument for a modified method of suturing in bypass surgery. 2 External view of instrument. 3 The instrument inserted in aorta sutured with blood vessel.

Madhvendra Singh
E madhur.rathod@gmail.com
T +91 9840520316

Manish Rathore
E mkr23jun@gmail.com
T +91 9884939303

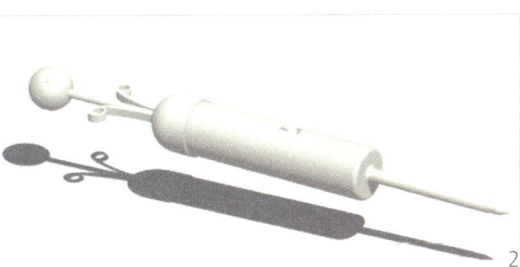

2

3

MANISH RATHORE AND MADHVENDRA SINGH

— IIT Madras —

Manish Rathore
I have always tried to seek opportunities to work on problems related with mankind. I believe that design is an outcome of continuous thinking. Working on this project has helped me in honing my designing skills. Extreme sports interest me.

Madhvendra Singh
I enjoy designing and developing products that improve our quality of life. At IIT Madras, I have been learning the concepts and tools that will make me an effective design engineer. I am excited at this opportunity of applying the knowledge that I have acquired into solving a practical problem. My hobbies include dancing, composing songs and meditation.

BURRA V.L.N. PRANEETH AND ABHISHEK GANDHI

IIT Madras

Burra V.L.N. Praneeth
Having learnt a good number of concepts in design synthesis and design analysis for the past two years, I made my first mark through this product. I believe that my interest in exploring new domains of design challenges is my main strength. I hope my remaining years at IIT Madras will give me all the inputs to manifest my creativity.

Abhishek Gandhi
Design has always enthused me. It goes back as far as solving design problems with Lego sets. Naturally, choosing a degree in design was the way forward. In a short span of two years at IIT Madras, I have caught on to the basics of product design. I am keen to further develop my appreciation in all aspects of design in years to come.

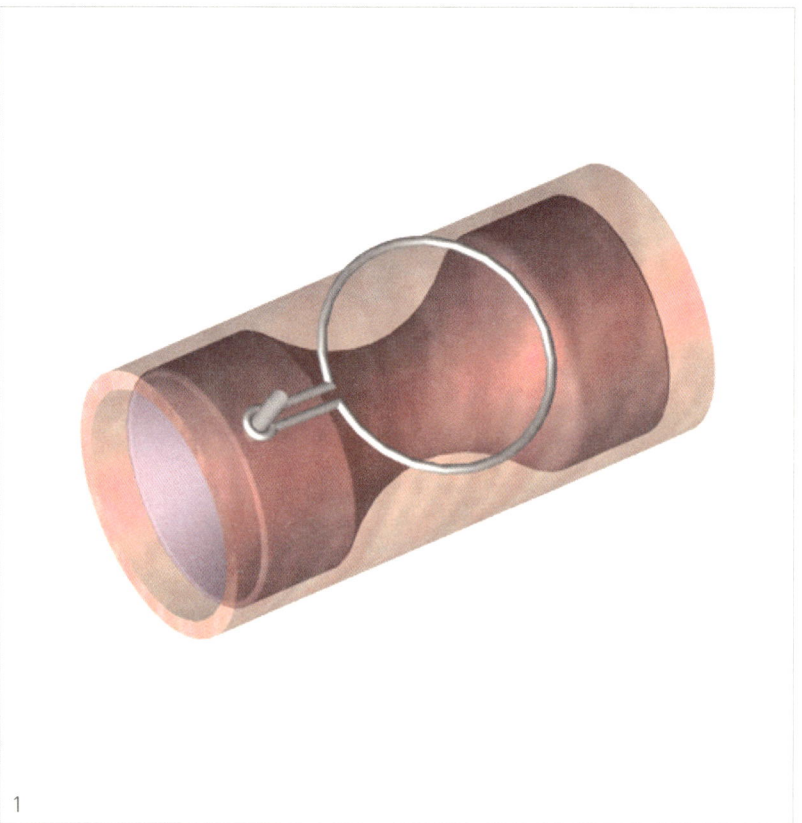

1

In Coronary Artery Bypass Graft surgery, it is necessary to occlude the aorta in order to provide the surgeon with a portion of aortic surface free from blood flow to implant the grafts. The use of the cardiovascular clamps makes it very cumbersome to operate and involves the risk of losing the grip.

Our design involves inserting an inflatable annular balloon into the aorta through a small hole by means of a guide wire. The profile of the balloon is such that when inflated, it will seal the walls of the aorta at the required place from the flow of blood. A template is used to locate the region in which the aortic surface needs to be cut and the graft is attached to the aorta in this region.

1 An apparatus for partial aortic occlusion during Coronary Artery Bypass Graft Surgery. 3 Pro/ENGINEER was used to model the design. Solid Surface Modelling techniques, scene rendering and editing functions were utilised to replicate the design. 4 The template helps in locating the operating area on the surface of aorta.

Burra V.L.N. Praneeth
E praneethbvln@gmail.com
T +91-9840131214

Abhishek Gandhi|
E abhishek.iitm@gmail.com
T +91-9840131214

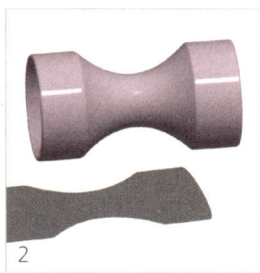

2

3

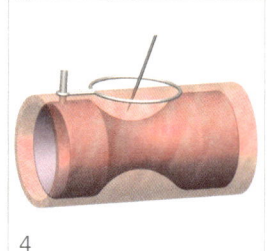

4

DEPARTMENT OF ENGINEERING DESIGN
INDIAN INSTITUTE OF TECHNOLOGY MADRAS

The primary objective of the Indian Institutes of technology is to provide scientists and technologists of the highest calibre, who engage in research, design and development to help progress our modern society towards self-reliance in its technological needs.

The faculty and students of the Institute at Madras are undertaking advanced research in science and technology, and providing cutting edge innovations, techniques and solutions to a wide range of industries and respective corporations. The Institute also provides knowledge-based technological services to satisfy the needs of the wider society. Life as part of IIT Madras is designed to stimulate students professionally, culturally and physically.

The Department of Engineering Design has state-of-the-art facilities for teaching and research including modelling and graphic art studios, and modern garage, a Computer Aided Engineering Centre. In addition, the department features ergonomics, mechatronics, product design, IC engines, controls and vehicle dynamics laboratories. These enable virtual prototyping, which has become the norm in many industries today, drastically reducing the time to market for new technologies. With the faculty collaborating with industry in research projects, teaching constantly reflects the advances in global best practices.

As part of their curriculum, students are introduced to the design process in the first year along with fundamental concepts in mathematics, science and engineering, graphic art, design aesthetics and art appreciation. In addition to being trained on the mechanical aspects of design they are also taught concepts in electronics, controls and embedded systems to obtain a well rounded set of skills.

Computers have great role to play in contemporary design practice. With modern CAD and CAM laboratories and courses in geometric modelling, finite elements and computational fluid dynamics, the curriculum provides the appropriate blend of theory and practice. It also includes courses in business management, economics, one foreign language, and the legal aspects of design such as intellectual property rights. Eight comprehensive courses during the fourth and the fifth year take students through all aspects of automotive design, from structures to electronics.

FURTHER INFORMATION

COURSE DESCRIPTIONS

SCHOOL OF ENGINEERING AND DESIGN

The Brunel School of Engineering and Design is one of the largest and most successful Design and Engineering Schools in the UK with international reputation for teaching and research. Our reputation in educating engineers and designers stems from the relevance of our undergraduate and postgraduate programmes to industry and the community at large.

Our approach to educating students is through the practical application of knowledge which promotes the development of creative skills and technical expertise. A significant content of industrially sponsored project based work, carried out both individually and in groups, and periods of industrial training ensure that our graduates attain the necessary leadership and communication skills that make them highly employable.

The projects exhibited at MADE IN BRUNEL 2007 demonstrate the talent and ability of our students to generate ideas and concepts and apply design principles to drive innovation and wealth creation, and we are extremely proud to be part of this.

Professor Savvas Tassou
Head of School of Engineering and Design

DESIGN

The family of Brunel Design degree courses now allows young design thinkers to explore and develop their creative potential, build competence in technological understanding and apply business knowledge to new product development. We have, over the last two decades, created a wealth of knowledge and our design programmes constantly evolve through close working relationships with commercial specialists across the globe. Our young Brunel Design graduates have the most comprehensive portfolios of transferrable skills.

Our strong relationships with manufacturing and service companies across the globe enable us to align our teaching to the constantly evolving needs of industry. All the courses benefit from real projects set by industrial partners and over 90% of our undergraduate students have gained professional and cultural experience through their year working abroad and attached to industrial placements. Many of our graduating students have already been offered jobs with the companies where they worked on placement; the endorsement of the quality of their work and the value of the industrial placement scheme.

Brunel Design represents the very best of creative thinking with innovative use of new technologies, new materials and excting new manufacturing processes. Our philosophy of innovation through realisation shows throughout this book. At the heart of all of our work is the fundamental need to design innovative new products with empathy for the targeted users. We design for people and so many of the exciting ideas within this book have been initially generated by users, developed together and realised appropriately. Brunel Design represents the synergy of all these aspects of innovation and this MADE IN BRUNEL book contains the very best design practitioners, each making their mark through their excellent work.

Design courses available:
- Industrial Design and Technology BA
- Industrial Design BSc
- Product Design BSc
- Product Design Engineering BSc
- Virtual Product Design BSc

Admissions Tutor:
Stephen Green
stephen.green@brunel.ac.uk

ELECTRONIC AND COMPUTER ENGINEERING

The subject area of Electronic and Computer Engineering encompasses a broad variety of courses that focus on today's connected digital society. We produce engineers, technologists and designers that are the architects and implementers of this technological revolution.

Our courses have strong links with industry; students can expect to conduct 'live' assignments in conjunction with industrial partners that set briefs during their studies. Many students choose to embark on a paid placement and have worked at companies like Walt Disney, Microsoft, Avid, Xerox, Dare Digital and many others. It is not unusual for graduating students to gain their first job with their placement employer and our record for this is one of the best in the University.

MADE IN BRUNEL features a range of engineering in society, multimedia technology and multimedia design projects. Typical project areas would include; including Intelligent Buildings, Videographics, Encryption and Security, Social Networking, Internet and Mobile services, Web Design, Photography, 3d Modelling for games and film and Sustainable Energy Management.

The areas of study within the subject area including courses in Multimedia Technology and Design, Broadcast Media Design and Technology and an extensive range of electronic and computer engineering BEng and MEng programmes.

Electronic and Computer Engineering courses available:
• Broadcast Media (Design and Technology) BSc
• Communication Networks Engineering BEng/MEng
• Computer Systems Engineering BEng/MEng
• Electrical Engineering with Sustainable Power Systems Management MEng
• Electronic and Computer Engineering MEng
• Electronic and Electrical Engineering BEng/MEng
• Electronic and Microelectronic Engineering BEng
• Internet Engineering BEng
• Mobile Computing BSc
• Multimedia Technology and Design BSc
• Networked Media Engineering BEng/MEng
• Space Engineering BEng/MEng

Admissions Tutor:
Prof John Stapleton
john.stapleton@brunel.ac.uk

MECHANICAL ENGINEERING

Engineering plays a key role within modern society in generating technological innovations and providing exciting employment opportunities. The mechanical engineering students who feature in MADE IN BRUNEL this year demonstrate the industry-relevant skills, creativity and intellect required to succeed in their chosen profession.

The quality of project work on display from mechanical engineering, underlines our commitment to innovation and excellence. Projects include a race car, a solar powered light aircraft, a novel fuel-cell based power and heating system for a modern dwelling, and a human-powered potable water device. Some of these projects are team-based and highlight the benefits of co-operative working so important to industry.

Mechanical Engineering at Brunel not only offers core mechanical engineering programmes but also specialisms in aerospace, aeronautics, motorsport, automotive design, building services and recently aviation engineering with a pilot studies option.

Mechanical Engineering courses available:
• Aerospace Engineering BEng/MEng
• Aviation Engineering BEng/MEng
• Aviation Engineering with Pilot Studies BEng/MEng
• Civil Engineering with Sustainability BEng/MEng
• Mechanical Engineering (with Aeronautics, Automotive Design, Building Services) BEng/MEng
• Motorsport Engineering BEng/MEng

Admissions Tutor:
Petra Godwin
petra.godwin@brunel.ac.uk

ACKNOWLEDGEMENTS

We would like to thank everyone who has supported and contributed to MADE IN BRUNEL 2007

Our Sincere Thanks To:

Yasar Khan, Angel Investor Magazine
Emma Harding, BAA
Joe Hardman, BAA
Neil Harvey, Bombardier Transportation
Veronica Bingham, Bombardier Transportation
Katie Tozer, Business Design Centre
Lynne Tarling, Butcombe Brewery
Victoria Smend, Bosch
Helen Watkins, Bosch
Neil Paul, Cadogan Tate
Mary Walsh, Cadogan Tate
James Cooper, Dare Digital
Deloitte & Touche
Peter Bishop, Design for London
Richard Croft, Easynet
Eddie Grosvenor, ECS Decor
Marco Fainello, Ferrari
Victoria Lefroy, Fuse PR
Jay O'Connor, Fuse PR
Dame Mary Richardson, HSBC Global Education Trust
Paul Sinclair, HSBC Global Education Trust
Ekow Eshun, ICA
Stuart Harper, InFocus
Kyle Ranson, InFocus
Richard Campbell, Institution of Mechanical Engineers
Andrew Ives, Institution of Mechanical Engineers
Lightweight Medical
Ajay Kathrani, London Development Agency

Alex McIntosh, Make Your Mark
Mindy Wilson, Make Your Mark
Jamison Combs, Natural Balance Foods
Andrew Barratt, NHS
Roger Bowden, Niftylift
Martin Cross, Niftylift
Simon Maher, Niftylift
Eddie Picton, Nimbus Rose
Elaine Rough, Nuffield Press
Clive Grinyer, Orange/France Telecom
Marcus Fairs, Pecha Kucha
Francis Hodge, Publicity Projects
Brian Kingham, Reliance Security
Tessa Lydekker Reliance Security
Alex Shapiro, Republic PR
Christina Stocks, Republic PR
Louise Bird, Rocket Mailing
Bob Athwal, RWE npower
Mike Evans, RWE npower
John White, Show Presentation Services
Tom Broomby, Square Group
Jean Betts, The Good Eating Company
Samantha Thomas, The Good Eating Company
Joe Ferry, Virgin Atlantic
Emma Kenny, Virgin Atlantic
Julian Richardson, Virgin Atlantic
Charles Vine, Virgin Atlantic

Philip Austin
Dave Branfield
Tanya Budd
Natalie Crowther
Sam Davies
Matt Gentle
Sir Digby Jones
Pat Jordan
Andrew Lee
Angus Macdonald
Patrick Quayle

Our Special Thanks To:

Linda Ajam
Terry Andrews
Liz Annetts
Mark Atherton
Paul Barrett
Peter Bird
Stuart Bonney
Les Botfield
Prof. John Boult
Prof. Clive Butler
Diana Burtinshaw
Sandy Bryant
Kitty Chisholm
Steve Cockett
Natalie Cooper
Leon Cruickshank
Sue Curley
Tania Cutter
Carolynne Deakin
Ian Dear
Dr. Hua Dong
Lyn Edgecock
Stephen Gardner
Clive Gee
Prof. Joseph Giacomin
Kathy Goddard
Petra Godwin
Stephen Green
Janie Grover
Ruth Harper
Francesca Insole
Ron Jackson
Dr. Shaheena Janjuha-Jivraj

Katy Jenkins
Prof. Chris Jenks
Gareth Jones
Paul Josse
Ian Kay
Angelina Karpovich
Andrew Kershaw
Tom Kissack
Mary Liddell
Paul Mardell
Benjamin McCalla
Tony Morris
John Morse
Hossein Nili
Karin O'Neill
Linda Paul
Baljinder Ram
Alison Russell
Saysavangh Sith-Ratanavong
Prof. Abdul Sadka
Gordon Scott
Dr. Siva Sivaloganathan
Dave Snowden
Prof. Tadeusz Stolarski
Hema Tank
Prof. Savvas Tassou
Prof. Linda Thomas
Liz Thomas
Peter Turner
Paul Turnock
Gordon Williams
Tony Wood
Paul Worthington

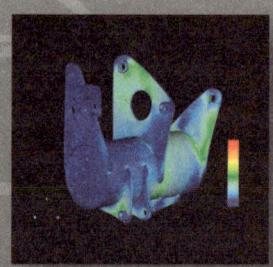

Everyone gets the same treatment

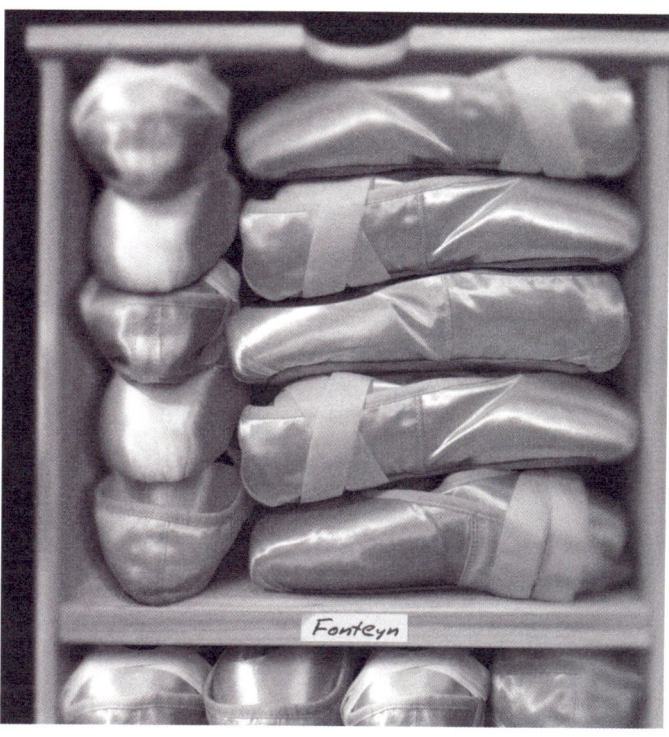

We care for everything we move and store. Whatever it is, if it matters to you it matters to us.

When you're moving our experienced estimators will make sure that they fully understand your needs before drawing up a comprehensive move-plan and quotation.

And if you need to store your property we will provide the most suitable and cost effective facility, whether it be containerised storage or a secure private room.

To find out how we can help you with your moving or storage needs please call us direct on 0800 008 6040 or find out more at www.cadogantate.com.

Cadogan Tate
Everything, handled with care

Moving Storage Shipping

Automated People Mover
Gatwick Airport

CITYFLO
Control Centre

Light Rail
Docklands Light Railway

Central Line
London Underground

Electrostar
Southern

Meridian
Midland Mainline

Turbostar
Chiltern Railways

Fleet Maintenance
Electrostar c2c

Automated People Mover
Stansted Airport

We Move London

Millions of people.
We move them every day.

We design, build, maintain and support
London's rail networks.
Metro cars on the London Underground.
An automated light rail system in Docklands.
People mover systems at London airports.
Diesel and electric multiple units that connect
London to the rest of the United Kingdom.

To ensure you arrive safely and on time,
our CITYFLO rail control solutions provide
a full range of automatic train control
technology from semi-automatic to fully
driverless systems.

If you ride the rails in London, chances are
you're doing it on a Bombardier product.
Effective solutions for public transport.
That's our business.

www.bombardier.com

BOMBARDIER

make **YOUr** mark

IN MANUFACTURING
& ENGINEERING

Make Your Mark is a national campaign aimed at inspiring young people aged 14-30 to be enterprising by having an idea and making it happen - in business start up, social enterprise or at work. Where better to do this than in the manufacturing and engineering industries; there are so many opportunities and you don't have to be a rocket scientist to make a difference
www.starttalkingideas.org/manufacturing

The School of Engineering and Design is currently based within the four 'towers' on campus. Tower A, pictured, is home to Brunel Design.

NOTES

◄ Preparing the BR-8 race car for the Formula
Student competition requires dedication,
perseverance and an eye for detail.

169

NOTES

◄ Students fully experience the 'hands-on' approach,
which develops and excellent understanding
of materials, as well as a keen technical skills
in product prototyping and manufacture.

171

NOTES

◄ The development process is often a group activity;
brainstorming and sharing new ideas an concepts to find
the optimum solution, as well as building prototypes for a
team to evaluate and improve upon collectively.

ISAMBARD
KINGDOM
BRUNEL
1806-1859

NOTES

◄ Isambard Kingdom Brunel was one of the most innovative minds of his age. He inspires this University, which dedicated this statue to him in recognition of his bicentenary, as well as the 40th anniversary of the university's Royal Charter in 2006.

NOTES

◄ BITLab is a state-of-the-art facility, which has promoted
innovation in multimedia, particularly with regard to
motion-capture techniques and three-dimensional imaging.

177

NOTES

INDEX